FIBER OPTIC SENSORS
PRINCIPLES AND APPLICATIONS

NIPA® GENX ELECTRONIC RESOURCES & SOLUTIONS P. LTD.
New Delhi-110 034

FIBER OPTIC SENSORS
PRINCIPLES AND APPLICATIONS

B D Gupta
Associate Professor
Department of Physics
Indian Institute of Technology Delhi

NIPA® GENX ELECTRONIC RESOURCES & SOLUTIONS P. LTD.
New Delhi-110 034

NIPA® GENX ELECTRONIC RESOURCES & SOLUTIONS P. LTD.

101,103, Vikas Surya Plaza, CU Block
L.S.C. Market, Pitam Pura, New Delhi-110 034
Ph : +91 11 27341616, 27341717, 27341718
E-mail: newindiapublishingagency@gmail.com
Web: www.nipabooks.com

For customer assistance, please contact
Phone: + 91-11-27 34 17 17
Fax: + 91-11- 27 34 16 16
E-Mail: feedbacks@nipabooks.com

ISBN: 978-81-19103-98-0

Composed and Designed by NIPA®.

PREFACE

The present book is an introduction to the rapidly emerging field of fiber optic sensors that is having significant impact upon areas such as guidance and control, structural monitoring, process control and medicine. The book will enable the reader to

- understand the benefits of fiber optic sensor technology
- know the basic principles upon which the fiber optic sensors rely
- become familiar with various physical, chemical and biological parameters that can be measured using fiber optic sensors
- understand how this technology is being used today and how it is likely to impact future systems and products.

The book is introductory and describes different kinds of sensors in a single book. The stress is on the principles and applications of various kinds of sensors. The book is intended for the postgraduate and Ph.D. students who would like to work in the field of fiber optic sensors. It will also provide extra knowledge to those who are doing courses on Optoelectronics and Fiber Optics. The scientists and engineers who are interested in learning about the operation and application of the fiber optic sensor technology will find this book very useful.

The first chapter of the book gives the introduction of the fiber optic sensors. The basic instrumentations required for fiber sensor are the optical fiber, light source and the detection system. In chapter 2 to 4 I describe briefly fiber optics, light sources and detectors respectively. Chapter 5 is devoted to the optical fiber sensors utilizing evanescent field absorption phenomenon. The sensors that do not require any reagent for absorption of light are discussed in this chapter. There are many chemical parameters that require the mediation of a reagent for their detection. Such kinds of sensors are described in chapter 6. The modulation of light in terms of intensity, phase, frequency, wavelength and polarization occurs in fiber optic sensors. Chapters 7 to 10 are devoted to fiber optic sensors based on wavelength, phase, polarization and frequency modulation, respectively. In chapters 11 and 12, sensors utilizing fiber Bragg grating and surface plasmon resonance phenomenon have been described. One of the most important advantages of optical fiber sensors is the distributed sensing. It enables the measurement of a physical or a chemical parameter as a function of position along the length of an optical fiber, and hence provides a unique capability to measure spatial variations of these quantities. Such systems are finding applications in industrial and environmental sensing. Chapter 13 is devoted to the distributed sensors. In the last chapter, simple, easy to make intensity modulated sensors have been described. The book should meet the requirements of students of science and engineering who are willing to work or are working in the area of fiber optic sensors.

I am grateful to my Teachers and Professor M.S. Sodha and Professor A.K. Ghatak who had encouraged me to work in the area of fiber optic sensors. Since the time I started work in the field of fiber optic sensors I had numerous stimulating discussions with my colleagues Professor B.P.Pal, Professor K.Thyagarajan, Professor Anurag Sharma, Professor Arun Kumar and Dr. M.R.Shenoy. I gratefully acknowledge them for these discussions. Writing a book in a short time was not possible without the help of many individuals. I am particularly grateful to my student Mr. Anuj Kumar Sharma for helping me in making figures and providing numerous suggestions towards the improvement of the book. My continuous interactions with my Ph.D. students have led to a deeper understanding of the field of fiber optic sensors. For this I would like to thank Dr. C.D.Singh, Dr. S.K.Khijwania, Dr. N.K.Sharma, Mr. Anuj K. Sharma and Mr. Rajan Jha. Finally, I owe a lot to my family, particularly to my wife Uma and my sons Amit and Puneet, for allowing me to spend long hours in writing and preparing this manuscript.

August, 2005 B.D. Gupta

Table of Contents

1

Optical Fiber Sensors

1.1 Introduction

Since 1990s optoelectronics and fiber optics technology underwent significant advances due to innovations in the telecommunications, semiconductor and consumer electronics sectors. The developments of compact disk (CD) players, laser printers and the high performance and reliable optical fiber communication are the result of these advances. The optical fiber sensor technology is a direct outgrowth of the revolutions taken place in optoelectronics and fiber optic communication industries. Many of the components associated with these industries are used for optical fiber sensor applications. In the beginning the main hurdle in the development of sensors was the cost of components such as optical sources and detectors. With the time, components prices are falling and the improvements in fiber, laser and photodetector technology have taken place. It is expected that as the time progresses the optical fiber sensors will replace the conventional devices for the measurements of various physical, chemical and biological parameters such as rotation, acceleration, electric and magnetic fields, temperature, pressure, acoustics, vibration, position, strain, humidity, viscosity, pH, glucose, heavy metals, gases, viral infection, pollutants *etc*.

Optical fiber sensor technology is an extremely promising and fast growing technology. At present there are about 100 companies with optical fiber sensor products available and 150 with products in development. It shows the utility of the technology and acts as a thrust for further development in this field. The current market for optical fiber sensor systems has been reported to be in the region of $550 million. The global market for sensors is expected to reach more than

$50 billion by 2008. The tremendous interest and activity in optical fiber sensors may be attributed to the number of advantages discussed in the next section.

1.2 Advantages of optical fiber sensors

The inherent advantages of optical fiber sensors over the conventional electronic sensors are the following:

1. ***Explosion proof:*** In optical fiber sensor, the primary signal is an optical. Therefore, there is no risk of spark or fire. Thus, it can be used in medical sciences without presenting any risk to patients. Further, these are safe for operation in hazardous environments such as oil refineries, grain bins, mines, and chemical processing plants.
2. ***Immune to radio frequency (RF) and electromagnetic (EM) interference:*** Since the fibers are composed of dielectrics such as glass or plastics, there is no need of electrical isolation of patients or elimination of conductive paths in high voltage environments. Further, the signal cannot be interfered by the static electricity of the human body. Optical fiber sensors can be used in electrically noisy environment, strong magnetic fields *etc*. without any interference of electromagnetic and radio frequency.
3. ***Environmental ruggedness and resistant:*** The optical fibers are manufactured from non-rusting materials such as plastics or glasses, therefore, the fibers have excellent stability when in permanent contact with electrolyte solutions, ionizing radiation etc. Further the fibers can withstand high temperature as high as 350°C. Special fibers can extend sensor operation beyond 350°C to as high as 1200°C.
4. ***Small size, light weight and flexible:*** The light guiding core of the fiber typically ranges from 5-600 μm in diameter. Further a 200 μm plastic clad silica fiber can be bent round a 1 cm radius mendrel. This implies that a sensor with a very small probe can be manufactured. This is advantageous, at least, in the case of minute sample and for invasive sensing in clinical chemistry and medicine. Further, due to light weight and small size these can be used in aircrafts.

5. ***Remote sensing:*** With the availability of low loss optical fibers, the optical signal can be transmitted up to a long distance (≈ 10-1000 m). Thus the remote sensing is possible with the optical fiber. This is important when the samples are hard to reach, dangerous, too hot or too cold, in harsh environments, or radioactive.
6. ***High sensitivity:*** The optical fiber sensors are highly sensitive and have large bandwidth. When multiplexed into arrays of sensors the large bandwidths of optical fibers themselves offer distinct advantages in their ability to transport the resultant data.
7. ***Potential of distributed sensing:*** Distributed sensing is one of the advantages where there is no competition. It enables the measurement of a physical or a chemical parameter as a function of position along the length of an optical fiber, and hence provides a unique capability to measure spatial variations of these quantities using a compact, inert and non-electrical sensing cable. Such systems are finding applications in industrial and environmental sensing.
8. ***Compactness:*** With the availability of solid-state configurations (small size sources and detectors) it is possible to design a compact optical fiber sensor system.

1.3 Generic optical fiber sensor

The block diagram of a generic optical fiber sensor is shown in Fig. 1.1. The basic instrumentation required for the sensor is light source, detection system, referencing scheme and sensor geometry. In the diagram photodetector 1 is used to check the light source output power fluctuation and hence is for referencing. The light from the source is coupled into the fiber using suitable optics. The sensing action occurs either within the fiber or external to the fiber. To communicate the change in output power taking place due to the measurand either the same or different optics can be used. The change in output power is detected by the photodetector 2. The instrumentation can be simple or complex. It depends on the particular application. The choice of light source, the characteristics of the optical fiber and the detection system used are critical in determining the device capabilities such as the sensitivity and dynamic range. For example, the material of the fiber determines the usable range of wavelengths. Since the performance of the sensor relies on the major components like optical fiber, source and the detector it is very essential to know the basics of fiber optics and the characteristics of the sources and the detectors which have been used and can be used for the design and development of the fiber optic sensors. In chapter 2, a review of the basic fiber optics is provided which should form as an introduction to the beginners. Chapters 3 and 4 have been devoted to the various optical sources and the detectors respectively.

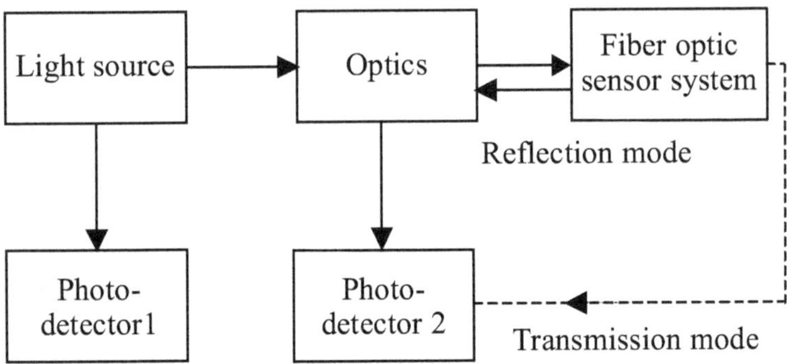

Figure 1.1: Block diagram of a generic optical fiber sensor.

1.4 Classifications

There are several ways in which optical fiber sensors for both physical and chemical parameters may be classified. One of the classifications is the following:

(i) External sensors,

(ii) Intrinsic sensors, and

(iii) Extrinsic sensors.

In the external sensors, the optical fiber used is bare-ended. It only guides and collects the light. There is no probe attached to its end or on the surface. The examples of external sensors are laser Doppler velocimeter, non-contact displacement or vibration sensors *etc*. In intrinsic sensors, the optical fiber becomes the active element of the sensor. The phenomenon or the parameter under investigation modifies the light propagation in the fiber. The optical fiber, in extrinsic sensor, is used to transmit the light from the source to an optical sensing head or probe attached at the other end of the fiber, and then returns the resultant light signal to a suitable detector. The fiber in extrinsic sensors is passive and plays no role in the sensing operation. It only provides a means by which light may be transmitted and received from the sensor, opening up regions normally inaccessible to optical measurements.

1.5 Modulation schemes

Optical fiber sensors may also be subdivided in accordance with the modulation scheme used. The divisions are as follows:

1. Intensity modulated sensors,
2. Phase modulated sensors,
3. Polarization modulated sensors,
4. Frequency modulated sensors, and
5. Wavelength modulated sensors

In intensity-modulated sensors, change in the intensity of the transmitted light is measured as a function of the perturbing environment. These sensors require more intense light source to function. These sensors are conceptually simple and cheap. The simplest example is a fiber embedded in a material as a crack sensor. As long as the fiber material is intact, there is no change in light output. If a crack develops in the fiber, it reduces light intensity or may cut the output light altogether. Such a sensor can measure structural integrity of a building, tunnel or bridge. The phase-modulated sensors are based on interference phenomenon. In this kind of sensors the phase of light in a sensing fiber is compared with the phase of light in a reference fiber. These sensors are more accurate than intensity modulated sensors and have large dynamic range. But they are more expensive than intensity modulated sensors. These sensors have few important limitations. One of these is that they require coherent light source. In addition, there is an inherent ambiguity because a delay of 360° produces the same effect as no delay or 720° delay. This requires a track of how many cycles of shifting occur during measurements. In the polarization modulation based sensors, a plane polarized light is launched in the fiber and the change in the state of polarization is measured as a function of the perturbing parameter of interest. A number of parameters can affect light polarization in a fiber for sensing applications. For example, magnetic field causes Faraday rotation of the plane polarized light by an angle proportional to the strength of the magnetic field. The frequency modulated sensors use Doppler effect while the wavelength modulated sensors are based on fluorescence phenomenon. Chapters 5 to 12 and 14 will be devoted to various physical and chemical sensors relying on these schemes. One of the most important advantages of optical fiber sensors is the distributed sensing. It enables the measurement of a physical or a chemical parameter as a function of position along the length of an optical fiber, and hence provides a unique capability to measure spatial variations of these quantities using a compact, inert and non-electrical sensing cable. Such systems are finding applications in industrial and environmental sensing. In chapter 13 we shall describe the technique of distributed sensing and few distributed sensors.

1.6 Fields of applications

Optical fiber sensors are finding their way into an ever-increasing number of applications as scientists, engineers and designers take advantage of their unique abilities. Their applications in some of the fields are described below.

1. ***Pollution monitoring (sea, water and air):*** Optical fiber sensors can be used to detect a large number of toxic substances and other chemical constituents such as formaldehyde, ammonia, nitrogen oxides, chloroform, hydrogen sulfide, sulfur dioxide, hydrocarbons *etc.* in sea, water and air. The main advantage of optical fiber sensors is rapid in situ measurement and remote sensing. Since the signal is optical rather than electrical, there will not be any electrical interference that are normally present in a typical shipboard environment.

2. ***Buildings:*** Optical fiber sensors also find application in structural monitoring. By measuring stress and strain in bridges, buildings, dams and tunnels – where failure can mean disaster – these sensors provide crucial analysis. The sensors are small enough to be embedded into the structure of buildings or vehicles and therefore can monitor passively and continuously the structure, providing information that can be tracked over time to indicate important structural changes. These sensors can provide early warning of weaknesses in structure so that minor repairs can be made before

major disaster strikes. In petroleum engineering, for example, these sensors are used for remote monitoring of the structure and function of an oil well to prevent hazards or malfunction. As the search for oil intensifies, wells are being drilled deeper, and monitoring equipment faces great demands in terms of pressure and temperature. Optical fiber sensors are also being used in squeezing the last ounce of production from old oil wells by optimizing the amount of water or steam injection used to efficiently remove reserves.

3. ***Biomedical sciences:*** The greatest field of application is sensing clinically and biochemically important analytes. Accurate and rapid measurement of the level of these analytes in blood are critical to good medical practice and patient care. To measure, blood samples are withdrawn from the patient and sent to the clinical laboratory to determine their contents. The delays and potential errors that can be introduced because the laboratory is far from the patient and from the physician who interprets the test results, may cause therapeutic decisions to be made without adequate information. If a large number of measurements are to be made per day then withdrawing sample and taking to the laboratory each time is not practical. Further, elderly, critically ill, or very small patients simply cannot afford to lose any blood. Fiber optic sensors with miniaturized probe can overcome these problems. It can be fitted into a catheter that can be inserted into blood vessels and hence analytes can be monitored continuously. The following measurements can be performed by optical fiber sensors:

(a) *In vivo* blood gas analysis (*i.e.* measurement of pH, pO_2, pCO_2).

(b) Monitoring of blood flow, blood pressure, hemoglobin concentration in blood.

(c) Total amount of protein, urea, glucose, cholesterol *etc.* in blood or urine.

(d) Drug monitoring.

(e) Detection of infection diseases

(f) Detection of various forms of cancer

(g) Monitoring of temperature and pressure of critical organs (such as brain)

4. ***Biotechnology:*** The main problem with conventional electrodes or electrode-based biosensors is the sterilization. The measurement of physical and chemical parameters in bioreactors requires sterilized sensors. For optical fiber sensors, sterilization is not a problem. The fibers can be sterilized by steam at 115-130°C without compromising their performance. Thus optical fibers can be used in sensing O_2, pH, pCO_2, glucose, glutamate in fermentation plants.

5. ***Ground water monitoring:*** Because of public concern about the quality of drinking water, continuous monitoring of ground water has become a necessity. By introducing fibers down to the ground water level one can monitor the pH of the water and the amount of chloride, uranium, organic pollutants and tracer substances present in water before digging a well. This will save drilling cost because fibers hold (1-2 cm diameter) allows the use of small bare-holes. Further, using optical fibers the quality of the water can be monitored in situ and in real time. Thus, there will not be any need to take sample and get it analysed in laboratory.

6. ***Process control:*** Production efficiency and product quality depend on process control. The optical fiber sensors can do on-line measurements and that in factory itself. Further there is no risk of fire/spark in the factory due to fiber sensors. Another advantage is they can be used in explosive environment. During process temperature, pressure, flow, liquid level, analytical parameters *etc* can be measured with optical fiber sensors.

7. ***Titrimetry:*** The optical fibers can be used in various titrimetric procedures including acid-base titrations, argentometry, bromometry and iodometry, complexometry, or redox titrations. The advantage is less cost because electrodes are expensive.

8. ***Military/Aerospace:*** Optical fibers can be used in gyroscope for rotation measurement, flight controls, engine monitoring, nuclear radiation testing, security systems *etc*.

9. ***Electrical Utilities/Power Plants:*** The temperature of various equipments (such as transformer) in power plants and the current and voltage in power stations can be sensed using optical fibers.

10. ***Automation/Machines:*** The position, vibration and strain can also be measured using optical fibers.

1.7 Issues

Optical fiber sensors have their limitations and these need to be appreciated if their advantages are to be exploited. Areas of concern include

1. ***Cost:*** The optical fiber sensors are costlier than the conventional electrical-based sensor systems. Therefore, optical fiber sensors should be used in situations where conventional sensor systems cannot be deployed.

2. ***Ambient light:*** In an optical fiber sensor ambient light can interfere with the signal of interest. Thus the measurements should be carried out in the dark. If it is not possible then the pulsed interrogating light should be used. Further the phase sensitive detection can be used to remove background light.

3. ***Response time:*** The response time of the sensor can be long, particularly in the case of multi-phase chemical sensing technique. This can be reduced by designing small probes.

4. ***Long-term stability:*** Optical fiber sensor is an emerging technology. There is little information available on the long-term performance, drift and reliability. The long-term stability of the sensor depends on the optical characteristics of the source, detector, optical fiber and the probe. A change in any of these can reduce the stability of the sensor.

5. ***Dynamic range:*** In the case of some of the fiber optic sensors, the operating dynamic range may be smaller than that of the electronic sensors.

6. ***Robustness:*** With the exception of polymeric fibers, optical fibers and sensors tend to be brittle, and adequate protection must be provided prior to their deployment. For example, for monitoring

concrete structures, the sensor system has to be chemically protected from the alkaline matrix and physically protected from potential damage from the aggregates.

7. ***Transduction:*** This is the efficiency with which the measurand of interest is transferred to the sensing region. Thus the measurand should be transferred efficiently from the substrate to the sensor. Various factors can influence this transfer.

8. ***Data interpretation:*** The interpretation of data from the sensors may not be straightforward because the output can be influenced by other parameters. Thus the effect of other parameters on the output should be controlled.

In addition to these, the other disadvantage of the optical fiber sensors is their unfamiliarity to the end user. However, with the time the situation is changing. The prices of optical sources, detectors and fibers are falling dramatically in the market. At the same time more sophisticated optical circuits are coming up. If all this continues then very soon fiber optic sensors will occupy more prominent position in the sensor market.

2

Basic Fiber Optics

2.1 Introduction

In a fiber optic sensor the heart of the sensor is an optical fiber. An optical fiber in its simplest form is a cylindrical symmetric structure consisting of a central 'core' glass/plastic of uniform refractive index and of diameter 4-600 μm. It is surrounded by a 'cladding' glass/plastic of slightly lower refractive index. To provide the mechanical and environmental protection to the fiber, the cladding is covered with an external plastic coating (see Fig. 2.1). The light in the fiber propagates by bouncing back and forth from the core-cladding interface. The design of the fiber optic sensor depends critically on the choice of optical fiber. Telecommunication fibers are inexpensive but may not be the right choice for sensors. Some of the fiber optic sensors dictate the use of application-specific optical fibers. In this chapter we shall first describe ray analysis for the light propagation in the fiber. The modes propagation, numerical aperture of the fiber, fiber characteristics and different kinds of optical fibers used for sensing will be described later.

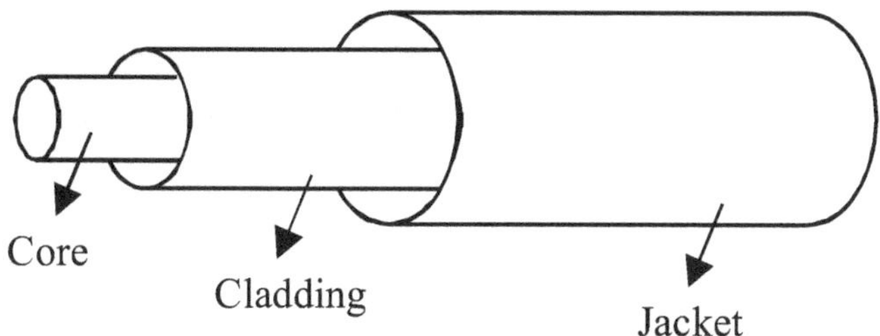

Figure 2.1: Geometry of an optical fiber.

2.2 Light propagation in an optical fiber

2.2.1 Ray propagation

To propagate the ray in a fiber, it must be injected from one of the ends of the fiber. The necessary conditions to inject such rays efficiently depend on the fiber structure as well as on the light source. To study the propagation of a ray in the fiber we use simple ray analysis. There are two kinds of rays that are propagated in the fiber. These are called meridional rays and skew rays. Rays that pass through the axis of the fiber are called meridional rays. There are a large number of rays that travel through the fiber without going through the axis of the fiber. These rays are called skew rays. The ray analysis of a meridional ray is simple and gives a close approximation of what actually takes place. The ray analysis of skew rays is cumbersome. In reality, it is nearly impossible to avoid the propagation of skew rays in the fiber.

Consider an interface between two optically different media of refractive indices n_1 and n_2. When a ray is incident at the interface both refraction and reflection take place (see Fig. 2.2). If θ_1 is the angle of incidence of the ray then according to Snell's law

$$\theta_1 = \theta_3 \tag{2.1}$$

and

$$\frac{\sin\theta_1}{\sin\theta_2} = \frac{n_2}{n_1} \tag{2.2}$$

where θ_2 and θ_3 are the angles of refraction and reflection, respectively. Thus, if a ray travels from a high refractive index medium (n_1) to a low refractive index medium (n_2) it bends away from the normal as shown in Fig.2.2. The reverse holds for the ray traveling from low to high refractive index medium. In Fig. 2.2 if angle θ_1 increases then the angle θ_2 also increases. When θ_2 becomes equal to 90°, the refracted ray does not travel through n_2 medium. Applying Snell's law for this case we obtain

$$\sin\theta_1 = \frac{n_2}{n_1}$$

The angle of incidence θ_1 for which $\theta_2 = 90°$ is called the critical angle, θ_c, and can be written as

$$\theta_c = \sin^{-1}\left(\frac{n_2}{n_1}\right) \tag{2.3}$$

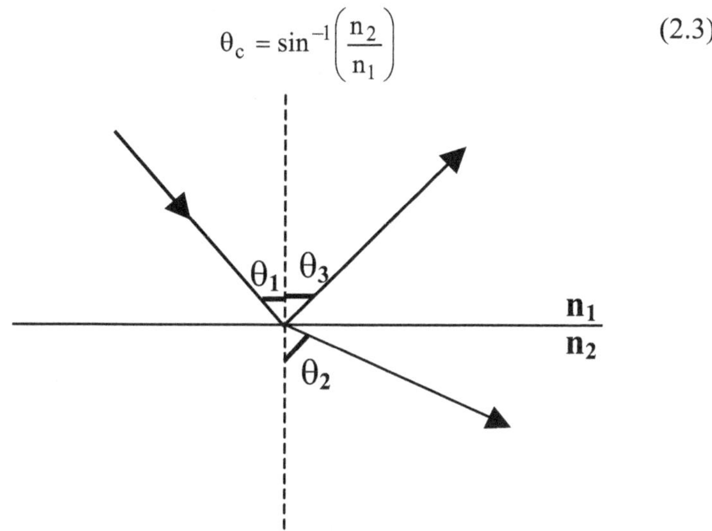

Figure 2.2: Incident, reflected and refracted rays for $n_2 < n_1$.

If the angle of incidence is greater than the critical angle, no refracted ray exists and the incident ray is totally reflected from the interface. This phenomenon where the incident ray is reflected back is called the total internal reflection. This can occur only when light travels from high refractive index medium to low refractive index medium.

Figure 2.3 shows the side view of an optical fiber with n_1 and n_2 as refractive indices of core and cladding respectively. If a ray launched in the fiber makes an angle greater than the critical angle at the core-cladding interface then the ray will suffer total internal reflection at the interface.

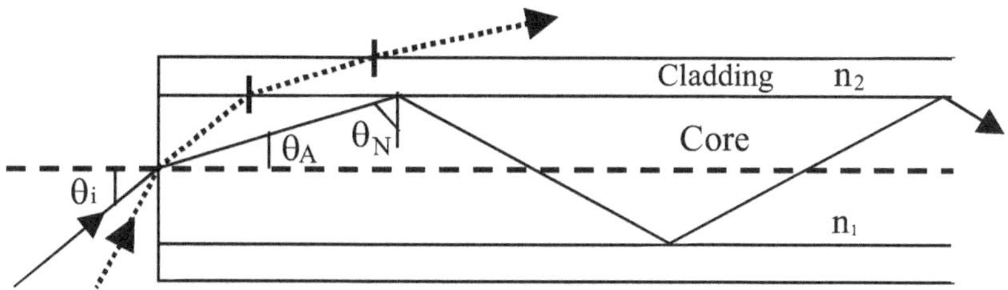

Figure 2.3: Propagation of rays in an optical fiber.

Due to the cylindrical symmetry of the structure, the ray would subsequently suffer repeated total internal reflections at the lower and upper interfaces. Hence this ray would get effectively trapped within the fiber as a guided ray to emerge at the other end of the fiber into air again. If the angle of the ray with the normal to the core-cladding interface is smaller than the critical angle the ray will exit from the interface and will not be guided. Thus for a ray to be guided

$$\theta_N > \theta_c = \sin^{-1}\left(\frac{n_2}{n_1}\right)$$

or,

$$\theta_A < 90^\circ - \theta_c.$$

Light guiding in fiber does not require a sharp change in refractive index at the core-cladding interface. A graded decrease in refractive index with distance from the fiber axis can also guide rays in the fiber. Such type of fibers are called graded index fibers. As light rays within the fiber's confinement angle move away from the fiber axis, they pass through layers of lower and lower refractive index which results in the bending of the rays toward the fiber axis as shown in Fig. 2.4.

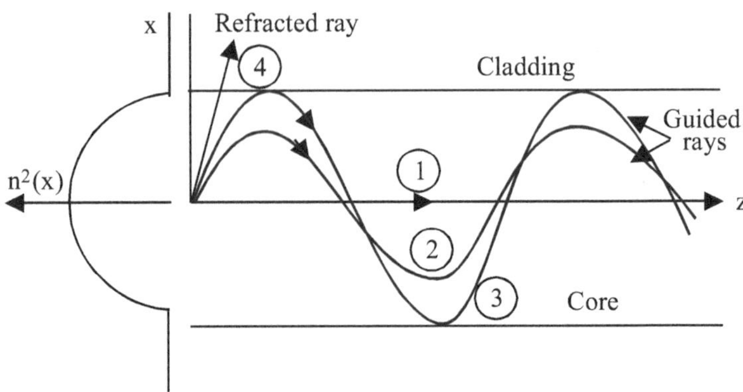

Figure 2.4: Propagation of rays in a graded index optical fiber.

2.2.2 *Mode propagation*

In the previous section propagation of light in optical fiber has been presented using the concept of total internal reflection. As we know, light is an electromagnetic wave, therefore a more rigorous analysis of wave propagation through optical fiber is required which is beyond the scope of this book. However, the rigorous analysis reveals that only rays corresponding to a discrete set of θ's (subject to certain electromagnetic boundary conditions imposed by the very nature of the fiber geometry) within the acceptance cone are guided by the fiber. All rays incident at the fiber input face at one of the angles θ of this discrete set form a so called mode of the structure. In terms of wave optics, a mode is defined as a specific electromagnetic field distribution that propagates through the fiber without any change in its amplitude distribution; the exact nature of the distribution is dictated by the boundary conditions imposed by the fiber itself through its refractive index distribution, core dimension, cladding refractive index and operating wavelength. Fiber parameters, namely core radius (ρ), relative core-cladding index difference $[(n_1-n_2)/n_1]$ and the wavelength of light (λ) together define a composite quantity, called V-number of the fiber whose value sort of decides the number of modes that can propagate in the fiber. The V-number is defined as

$$V = \frac{2\pi\rho}{\lambda}\left(n_1^{\,2} - n_2^{\,2}\right)^{1/2} \tag{2.4}$$

For a step index fiber (*i.e.* n_1 = constant) if V-number is less than 2.405 the fiber will support only one mode and it is called a single-mode fiber. On the other hand, if a fiber supports more than one mode (*i.e.* V>2.405) it is called a multimode fiber.

In terms of rays, the total number of modes propagating in the fiber increases as θ_c decreases or the relative refractive index difference $[(n_1-n_2)/n_1]$ increases. One can distinguish between higher order modes and lower order modes. Those rays which make angles θ_N close to the critical angle θ_c represent higher order modes while those with angles θ_N much higher than θ_c are called lower order modes.

2.3 Acceptance angle and numerical aperture

We return to Fig. 2.3 and investigate a particular condition, namely, what happens when θ_N equals θ_c. If the outside medium is air (n_{air} = 1) then we can write

$$\sin\theta_i = n_1 \sin\theta_A = n_1 \sin\left(\frac{\pi}{2} - \theta_N\right) = n_1 \cos\theta_N \tag{2.5}$$

For $\theta_N = \theta_c$, we obtain

$$\sin\theta_i = n_1 \cos\theta_c \tag{2.6}$$

Under this condition θ_i is the maximum angle of the ray with the fiber axis that is guided by the fiber. The angle $2\theta_i$ is called the acceptance angle of the fiber and the term $\sin\theta_i$ is called the numerical aperture (NA) of the fiber. In terms of refractive indices

$$\begin{aligned} NA &= n_1 \cos\theta_c = n_1 \left(1 - \sin^2\theta_c\right)^{1/2} \\ &= \left(n_1^2 - n_2^2\right)^{1/2} \end{aligned} \tag{2.7}$$

and the half acceptance angle θ_i is given by

$$\theta_i = \sin^{-1}(NA) = \sin^{-1}\left(n_1^2 - n_2^2\right)^{1/2} \tag{2.8}$$

A large NA represents a large acceptance angle and vice versa. Further, a large NA produces a large number of modes in the fiber (see Eq. 2.4).

2.4 Fiber characteristics

$\sin\theta_i = n_1 \cos\theta_c$

The characteristics of optical fibers depend on their material composition and the physical shape and size. Imperfections and microbends in the fiber affect the power loss and hence the fiber performance. In this section we shall concentrate on fiber losses and dispersion.

2.4.1 Fiber losses

Losses in fiber are expressed in decibels (dB). Decibels are logarithmic units defined as the ratio of output power to the input power. Thus

$$dB(loss) = -10\log\left(\frac{P_{out}}{P_{in}}\right) \tag{2.9}$$

The minus sign makes the loss a positive number. If input power is 1 mW and the output is 1 μW then loss is 30 dB. Decibels units are widely used because both losses and power levels are additive. The total loss of a series of components is the sum of their individual losses in decibels, and the loss of a length of fiber is the unit loss times the total distance.

There are three kinds of light losses in the fiber. These are material loss, light scattering and waveguide loss. Below we discuss them one by one.

Material loss

This is due to the absorption of light by the material of the fiber and the impurities present in the materials. However this loss is relatively small. The losses due to impurities can be reduced by improving the manufacturing processes of the fiber. The largest loss is caused by OH ions which cannot be sufficiently reduced. The OH^- impurity causes loss for particular wavelength bands. The worst loss occurs near λ = 1.4 μm. Figure 2.5 shows the fiber losses in the wavelength range 0.7 to 1.6 μm. The three absorption peaks in the curve are caused by OH^- ions.

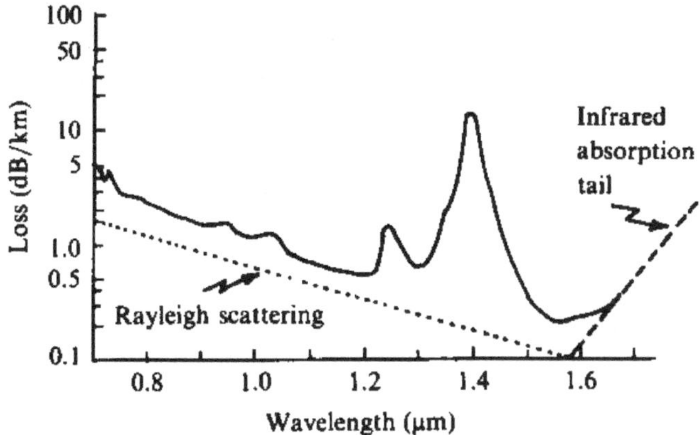

Figure 2.5: Losses in a silica fiber.

Light scattering

Light scattering occurs due to the density and hence the refractive index variations in the material of the core. If the refractive index variations are molecular in size, the power loss is due to Rayleigh scattering. These small variations cannot be eliminated. They scatter light in all directions. Rayleigh scattering loss varies as $1/\lambda^4$. Smaller the wavelength, larger the loss is. Figure 2.5 shows some of the effects of Rayleigh scattering losses.

Waveguide and bend losses

These losses are caused by imperfections and deformations of the fiber structure such as change in diameter and microbend (*i.e.* very minute disturbance in core size). These deformations change the angle of the ray at the core-cladding interface. If the angle becomes smaller than the critical angle the ray leaves the fiber and hence the loss occurs.

2.4.2 *Dispersion*

If a pulse of light is sent through a fiber it broadens in time as it propagates through the fiber. This phenomenon is known as pulse dispersion and occurs because of different times taken by different rays propagating through the fiber. In Fig. 2.6 two guided rays are shown. It can be seen that the ray making larger angle with the axis traverses a longer optical path length and takes a longer time to reach the output end of the fiber. Consequently, pulse broadens as it propagates through the fiber. Let us calculate the amount of dispersion in a step index fiber. In Fig.2.6 a ray making an angle θ_A with the axis covers the distance AB in time

$$t = \frac{AC + CB}{c/n_1} = n_1(AB)/c\cos\theta_A \tag{2.10}$$

where c/n_1 is the speed of light in fiber core and c is the speed of light in free space. For a fiber of length L, the time taken is

$$t = \frac{n_1 L}{c\cos\theta_A} \tag{2.11}$$

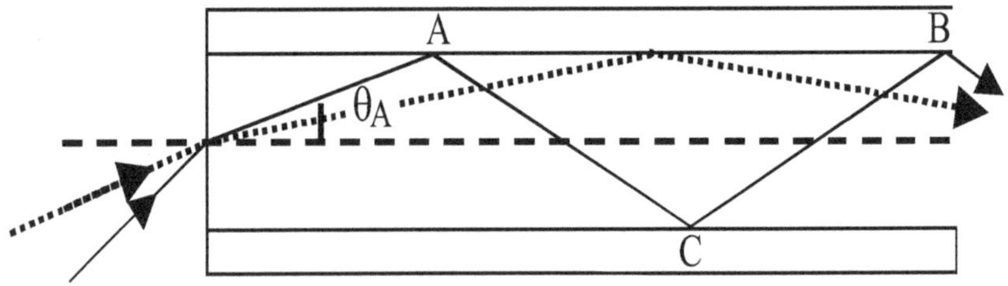

Figure 2.6: Propagation of two guided rays launched at two different angles.

The above equation shows that the time taken by a ray is a function of the angle θ_A made by the ray with the axis of the fiber. This gives rise to the pulse dispersion. For two extreme rays, $\theta_A = 0$ and $\theta_A = 90^o - \theta_c$, the time taken are

$$t(\theta_A = 0) = t_{min} = \frac{n_1 L}{c}$$

$$t(\theta_A = 90^o - \theta_c) = t_{max} = \frac{n_1^2 L}{n_2 c}$$

If all the guided rays are excited simultaneously in the fiber then the rays at the output end of the fiber would occupy a time interval of duration

$$T = t_{max} - t_{min} = \frac{n_1 L}{c}\left(\frac{n_1}{n_2} - 1\right) \quad (2.12)$$

For a fiber with $n_1 = 1.5$, $(n_1\text{-}n_2)/n_2 = 0.01$ and $L = 1$ km one obtains T = 50 ns. Thus an impulse after traversing 1 km through the fiber broadens to a pulse of duration 50 ns. This period of dispersion is called "*intermodal dispersion*".

In the case of parabolic graded index fiber, with some mathematics, Eq. (2.12) takes the following form

$$T = \frac{n_1 L}{2c}\left(\frac{n_1}{n_2} - 1\right)^2 \quad (2.13)$$

After substituting the values, one obtains T = 0.25 ns. The small value of T for a parabolic index fiber is because of nearly the same optical path length traversed by all rays along the fiber.

The pulse broadening also occurs due to the wavelength dependence of the refractive index of the material. This kind of dispersion is called "*intramodal dispersion*". Sometimes it is called "*chromatic or material dispersion*". Each light source has a certain spectral width $\Delta\lambda$. The pulse dispersion for this width is given by

$$T_m = -\frac{L}{c\lambda}\left(\lambda^2 \frac{d^2 n}{d\lambda^2}\right)\Delta\lambda \quad (2.14)$$

Dispersion can also occur even if the refractive indices of the core and cladding are strictly independent of the wavelength and only one mode is excited in the fiber. This is due to the intrinsic characteristics of the waveguide and hence is called “waveguide dispersion”.

2.5 Types of optical fibers

Basically, there are three kinds of optical fibers in use, namely the multimode step-index, multimode graded-index and the single mode step-index fibers. Their geometry and the refractive index profile are shown in Fig.2.7. In multimode step-index fiber, the refractive indices of the core and the cladding change in the form of a step function at the core-cladding interface. The core diameter is typically in the 50 μm to 600 μm range.

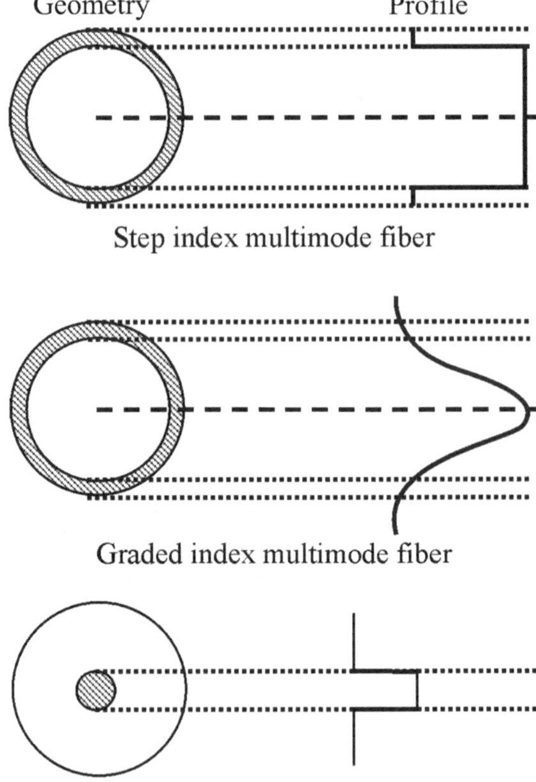

Figure 2.7: Geometry and refractive index profile for three different kinds of fibers.

These fibers have a comparatively simple manufacturing process and easy to work because of large core diameter. In the multimode graded-index fiber the refractive index of the core is maximum on the axis and decreases as we move away from the axis and becomes constant in the cladding. The dispersion is considerably reduced in the graded index fiber as discussed above. The fabrication of graded-index fiber is more complex. It is more difficult to control the refractive index well enough to produce accurately the variations needed for the desired index profile. The single mode fiber has a step index profile, but is distinguished by its much smaller core diameter and a smaller difference in refractive indices of core and cladding. As mentioned earlier for these fibers, the normalized frequency *i.e.* the V-value is less than 2.405. Typical core diameter of a single mode fiber is between 4 and 10 μm. The NA of the single-mode fiber is about 0.1 while in multimode fibers it could be 0.2-0.5 or even larger. Handling and fabrication of single mode fibers is more difficult.

2.6 Optical fibers for sensors

As mentioned in the beginning, the design of a fiber-optic sensor depends critically on the choice of optical fiber. Silica-based telecommunication fiber is inexpensive and readily available but may not be the right choice for all kinds of sensors. This is because of its marginal performance. Consequently, attention is now being given to the design of application-specific optical fibers. Some of these are described below.

2.6.1 Plastic-clad silica fibers

Plastic-clad silica (PCS) multimode fibers have been used mainly for laser power transmission and intensity modulated sensors. These fibers are step-index and have a pure silica core and a plastic cladding such as silicone rubber which can easily be removed mechanically or chemically. Thus the fiber can be used as intrinsic sensor. The unclad fiber can be used for sensing temperature, refractive index of the fluid, absorption *etc*. Sometimes a dye is immobilized on the surface of the core to sense a number of chemicals. PCS fibers can also be used for extrinsic sensors to transmit and collect light from the probe attached at the other end of the fiber.

2.6.2 *Fibers with dopants*

The properties of silica fibers can be modified by incorporating suitable dopants such as rare earths (Nd, Eu, Dy *etc*) or transition metal ions into the core of the fiber. The addition of these ions increases the acousto-optic, magneto-optic, non-linear and electro-optic coefficients which are small in pure silica. These effects are exploited to construct various fiber sensors for physical measurements or to improve the measurement region of the sensor. The rare earth doped fibers can also be used for temperature sensing. These fibers are also useful for chemical sensing based on absorption and fluorescence phenomena. Plastic fibers with organic fluorescent materials as dopant in the cladding have also been used for sensing.

2.6.3 *Infra-red Fibers*

For the detection of gases or chemicals which have absorption peaks in infra-red regions the silica-based optical fibers are not suitable. This is because the silica-based fibers have transparency in a limited infra-red spectral range and thus a limited system spectral response. Recently infra-red transparent fibers have been developed. These are fluoride glass, chalcogenide glass and silver halides-based fibers. The transmission window of the silica fiber is limited only up to about 3 μm to infra-red side, whereas those of fluoride, chalcogenide and halides fibers extend up to 7, 11 and 20 μm, respectively. Thus these fibers are most suitable for infra-red absorption spectroscopy.

Chalcogenide fibers have excellent chemical durability and glass stability. Two kinds of chalcogenide fibers have been fabricated. Sulphide based which transmit from 1 to 6 μm and telluride based for 3 to 11 μm region. The fiber design is step-index, with typical numerical aperture ranging from 0.2 to 0.4 depending on the core and cladding composition. The fiber dimensions vary from application to application. The dimensions typically range from 130/220 to 300/500 μm with a protective Teflon outer coating. Chalcogenide fibers can be bent without breakage to a radius of about 1 cm.

Halide glasses such as AgBr, KCl, CsBr and $ZnCl_2$ have an extended wavelength range up to 15-20 μm. The disadvantages with these fibers are that they have a high attenuation, are sensitive to humidity and are costly.

2.6.4 *Metal coated fibers*

To sense magnetic field, metal coated fibers have been developed. In these fibers a thin film of metal is deposited by electron beam evaporation technique onto bare fiber to uniform thickness. The metals mostly used are nickel and aluminium. The sensitivity of these fibers increases with the increase in the film thickness. The cross-section of a metal coated fiber is shown in Fig.2.8.

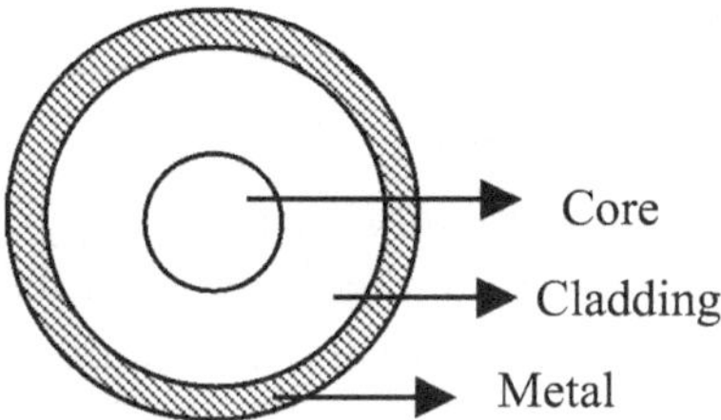

Figure 2.8: Metal coated fiber cross-section.

2.6.5 *Liquid-core fibers*

Liquids have been used as core materials for fibers where the special characteristics of the liquid have been required for a specific sensor purpose. The refractive index, scattering properties or Kerr coefficients of the core material are tailored to a particular application through the use of a specially chosen liquid. For example, a fiber filled with an organic liquid with significant scattering properties can be used as the basis of a distributed temperature sensor. Further, the fiber core filled with Kerr liquid can be used to measure the high voltage. However, problems exist in the fabrication of long lengths of such fiber and possible hazards from spillage of the core material. Additionally, such fibers are expensive to fabricate and are not widely acceptable in industrial installations.

2.6.6 *D-shaped fibers*

The D-shaped fibers are produced by polishing a flat on the fiber preform to give a D-shaped cross-section, and then the preform is subsequently pulled into fiber. The cross-section of the D-shaped fiber is shown in Fig. 2.9. Most of the cladding from one side of the fiber is removed and the thickness of the cladding material that is left is very small. The light modulation is achieved at this side. The main potential use of single mode D-shaped fiber is in chemical and distributed sensing.

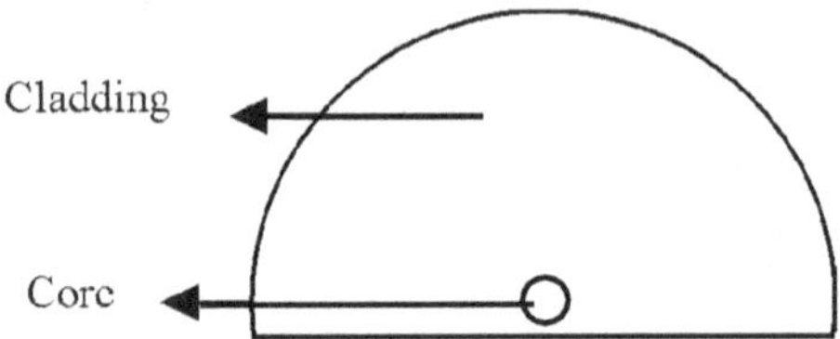

Figure 2.9: Cross-section of D-shaped optical fiber.

2.6.7 Polarization maintaining fibers

Many optical fiber sensors rely on the use of polarization effects. Therefore the maintenance of the state of polarization of light propagating in the fiber is important. Standard nominally circular single-mode fiber propagates two independent modes having orthogonal polarization states and having the same propagation constants. The fiber maintains input polarization only up to a short distance because depolarizing perturbations such as minor stress due to bends, twists and kinks, and changes in ambient conditions couple energy between the two subsequent polarization modes. Thus, if a linearly polarized light is incident in a nominally circular core fiber it will quickly become elliptically polarized after propagating over a short distance of the fiber. In order to overcome this polarization instability problem, polarization maintaining single-mode fibers are developed. These are called elliptical core, elliptical clad, stress lobe, and bow tie fibers. The geometrical cross-sections of these polarization maintaining fibers are shown in Fig.2.10.

2.6.8 Polarizing fibers

Polarizing single-mode fibers are those which guide only one direction of polarization. In polarizing fibers one of the two modes is severely suppressed or attenuated so that only one polarization mode propagates in the fiber. Various fiber designs have been investigated for polarizing fibers. The undesired polarization mode is suppressed by incorporating a large stress birefringence or by utilizing a depressed cladding or an elliptical core. For certain optimum operation, the fiber is required to be bent to a certain radius. These fibers are useful for gyroscopes.

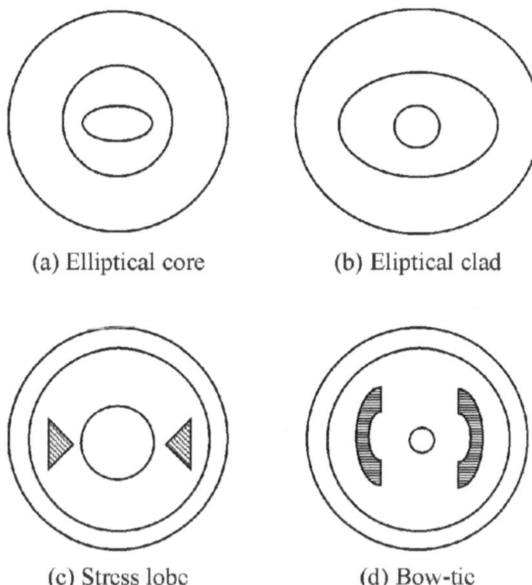

Figure 2.10: Cross-sections of polarization maintaining optical fibers.

2.6.9 *Circularly birefringent fibers*

In the circularly birefringent fiber a round fiber is twisted to produce a difference between the propagation constants of the clockwise and anticlockwise circularly polarized modes. These fibers are very suitable for magnetic field sensing.

2.6.10 *Holey fibers*

Holey or "microstructured" fibers have generated a wave of excitement because they promise properties that cannot be achieved in conventional optical fibers. In conventional fibers, light-guiding properties depend on the properties of the material used for its fabrication. Internal microstructures add another degree of freedom in controlling the light guiding in fibers. This opens new possibilities in designing sensors. The holey fibers guide light through air, avoiding the absorption, nonlinear properties, and material dispersion which occur in solids. Holey fibers are based on the concept of the photonic bandgap, a microstructure made of alternating layers of two materials with different refractive indices. This microstructure acts like a thin-film interference coating which selectively reflect or reject certain wavelengths. The reflected or rejected wavelengths depend on the thickness and refractive indices of the layers. In holey fibers these two materials are transparent solid like glass and the other is the air. The air is contained in the holes that run up to the length of the fiber. The holey fiber is made by stacking tubes and solid rods together. This bundle is then fused and fiber is drawn. The size and spacing of holes characterize the microstructure.

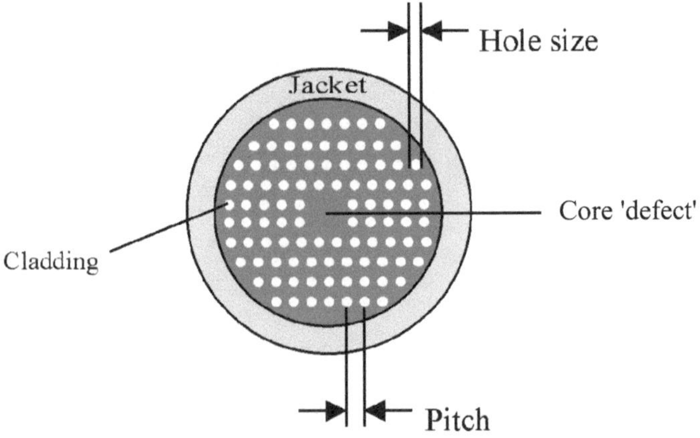

Figure 2.11: Photonic-crystal fiber.

There are two kinds of holey fibers depending on whether the light is guided through a solid core or a hollow region. But the light guiding properties in both the cases depend on the size and spacing of holes and the refractive index of the solid region. The fibers with solid core surrounded by a cladding layer containing holes running the length of the fiber are called photonic crystal fibers (Fig. 2.11). The cladding is a photonic bandgap material with average refractive index less than that of the solid core. Another type of holey fibers, called photonic-bandgap fibers, have a hollow core surrounded by a photonic-bandgap cladding. In these fibers, light is guided through the air filled core (Fig. 2.12). Holey fibers have recently been used for gas sensing.

2.7 Fiber selection for sensors

The performance of a fiber optic sensor highly depends on the selection of an optical fiber. The designer must take into account the factors such as operating wavelength, attenuation, power capability, size constraints, deployment condition, environment and the cost before selecting the fiber for the sensor.

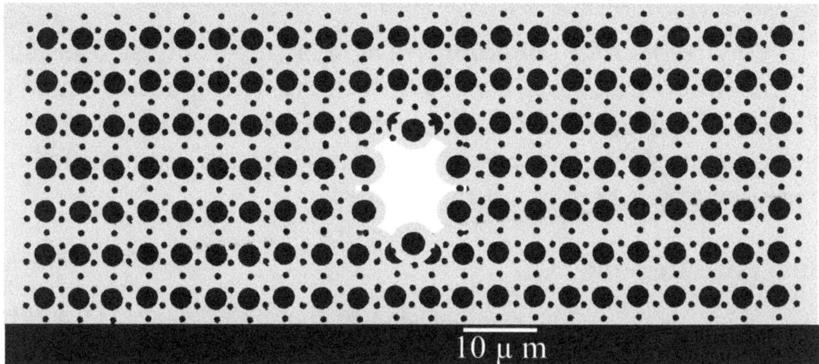

Figure 2.12: Photonic bandgap fiber.

Operating wavelength: The operating wavelength of the fiber is important for sensors that are using single mode fibers. For sensor's performance shot-noise limit and sensitivity are important. The shot-noise limit improves with increase in wavelength whereas sensitivity increases with decrease in wavelength. Shot-noise decreases because attenuation is low at higher wavelengths and hence the detected power is more. Because of opposite wavelength effect, it is difficult to choose an ideal operating wavelength on the basis of fibers used for sensors. The choice of operating wavelength generally depends on the availability of the light source, its output power and spectral characteristics.

Attenuation: If the length of the fiber used for sensor is small, which is indeed the case in most of the sensors, then the fiber attenuation is not a critical parameter.

Power capability: It is the ability of the fiber to efficiently accept and transmit optical power up to the other end of the fiber. This is important for intensity-modulated sensors. Multimode optical fibers with large core diameters and high numerical apertures are used for these kinds of sensors.

Space constraint: In interferometric sensors such as fiber gyroscope and hydrophone, the increase in length of the fiber increases sensitivity. If space is a constraint and such type of sensors are to be packed in that space then fibers of longer lengths and reduced sizes are important. Reduced-size fiber also reduces mechanical stress due to bending. This improves the overall reliability of the sensor.

Environmental conditions: The choice of optical fiber also depends on environmental conditions such as temperature and humidity. These conditions affect the fiber-coating materials and their characteristics. Special coating materials may be required for the operation of sensor at high temperatures.

Suggested Reading

1. M.J. Adams (1981) An introduction to optical waveguides. John Wiley & Sons, New York.
2. P.K. Cheo (1990) Fiber optics and optoelectronics. 2nd Edition. Prentice-Hall International, New Jersey.
3. A.C. Cherin (1983) An introduction to optical fibers. McGraw-Hill International, Japan.
4. A.K. Ghatak and K. Thyagarajan (2002) Introduction to Fiber Optics. Cambridge University Press.
5. A.K. Ghatak and K.Thyagarajan (1989) Optical electronics. Cambridge University Press.
6. Y. Suematsu and K. Iga (1982) Introduction to optical fiber communications. John Wiley & Sons, New York.

3

Optical Sources

3.1 Introduction

Optical sources are a key element in many fiber optic systems. This component converts the electrical signal into a corresponding light signal that can be injected into the fiber. Several factors such as wavelength, source output and spectral bandwidth are important when selecting light source to achieve the best performance of the sensor. The optical sources used for sensors range from hot filament lamps to semiconductor lasers. These sources can be placed into two groups - incoherent and coherent sources. The tungsten lamp and light-emitting diode (LED) are some of the incoherent sources while the lasers come in the category of coherent sources. The LED is normally preferred if the multimode fiber is used for the sensor. If the optical power requirement is high then the laser with multimode optical fiber is used. For the sensors based on interference phenomenon, coherent sources with single mode fibers are used. In the case of distributed sensing, short pulse high power lasers are used. In a laboratory demonstration experiment, a visible laser such as a simple He-Ne laser may be the preferred source while in commercial applications a compact and rugged source such as laser diode or an LED is preferred. Before using the source it is very essential to know its maximum power output, wavelength of emission, linewidth, radiation pattern and modulation bandwidth. The purpose of this chapter is to provide the physics of some of the light sources and their important operating characteristics that are used in sensors.

3.2 General requirements

For use in optical fiber sensor, an optical source must fulfill the following requirements:

1. The light intensity or the optical power output of the source must be large. In addition, the structure of the source must permit the effective coupling of light into the fiber. The power efficiency of a source is defined as

$$\eta_P = \frac{P_{out}}{P_{el}} \tag{3.1}$$

where P_{out} is the optical power emitted and P_{el} is the input electrical power given to the light source. The coupling efficiency is given by

$$\eta_c = \frac{P_c}{P_{out}} \tag{3.2}$$

where P_c is the power coupled into the fiber. The coupling efficiency depends on the structure of the source, its intensity distribution, fiber core diameter and numerical aperture. Increase in numerical aperture of the fiber increases the coupling efficiency. For a single mode fiber with core diameter in the range 3-10 μm, special high intensity and small-area source is required. Laser is the most suitable source for the single mode fiber. The total efficiency of the device or the system efficiency is given by

$$\eta_T = \eta_P \eta_c = \frac{P_c}{P_{el}} \tag{3.3}$$

The total efficiency must be large for the best performance of the sensor.

2. The wavelength of the source must be compatible with the wavelength most suitable for the sensor. With the broadband source one may use an interference filter of required wavelength or a monochromator if needed.

3. There should not be any fluctuation in output power of the source. Further it should not depend on temperature and other environmental conditions. If it depends then these parameters should be controlled.

4. For the fiber optic sensor to compete with conventional sensors, the light source used in a sensor should be comparatively cheap and highly reliable.

3.3 Light sources

There are two kinds of light sources, thermal and non-thermal. We shall describe them separately.

3.3.1 Thermal sources

In thermal sources, radiation is produced by heating a filament through the application of an appropriate electric current. The thermal sources are frequently classified with reference to their emission as a blackbody radiator. The power radiated by thermal sources is a function of their temperature. The primary law governing radiation is the Planck's law of radiation which gives the intensity of radiation emitted by a blackbody as a function of wavelength for a fixed temperature. The peak occurs at some wavelength. It shifts to shorter wavelengths for higher temperatures. The spectral radiance, R, of a black-body at temperature, T, in Kelvin, is given by

$$R(\lambda, T) = \frac{2hc^2}{\lambda^5 \exp\left(\frac{hc}{\lambda kT} - 1\right)} \tag{3.4}$$

where λ is the wavelength of radiation, h is Planck's constant, k is Boltzmann's constant and c is the velocity of light.

A plot of spectral radiance with λ at different temperatures is shown in Fig. 3.1. It can be seen that as the temperature increases, the total relative radiant power emitted increases and the peak of the output shifts to a region of shorter wavelengths.

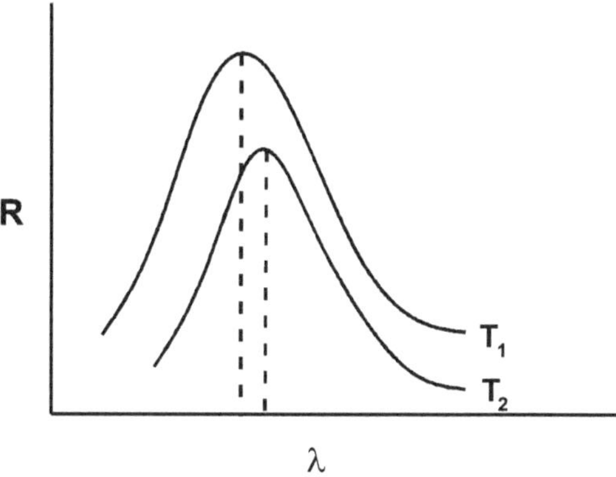

Figure 3.1: Spectral radiance as a function of wavelength, l, at different temperatures.

Different thermal sources operate at different temperatures and hence have different peak wavelengths. For example, in a tungsten filament lamp, described below, the filament is heated to approximately 2500°C. This is the maximum temperature that a tungsten filament can stand without evaporating quickly. Such a filament emits radiation in visible and infrared region but a greater fraction of radiation emitted is in the infrared region of the electromagnetic spectrum. The filaments used in thermal sources do not have unit emissivity, their emission can be very well approximated by a black-body. The thermal sources are incoherent sources.

Tungsten filament lamp

It is the most popular light source for optical instrumentation designed to use continuous radiation in the visible and the infrared region. The lamp is available in a wide variety of filament configurations and shapes of the bulb and the base. The filament is in coil or ribbon form. Light is produced by heating the filament of the lamp by applying a voltage across it. Radiation arises from the de-excitation of the atoms or molecules of the material after they have been thermally excited. The bulb is usually a glass envelope. For high temperature operation, instead of glass, quartz is used. During operation tungsten gradually evaporates from the filament and deposits on the inner surface of the bulb. This produces a dark film on the surface that decreases the output flux. This process also weakens the filament and increases its electrical resistance. The evaporation can be slowed down if an inert gas such as nitrogen or argon is introduced in the bulb. Recently this problem has been minimized by the addition of a halogen vapor (iodine, bromine) to the gas. The halogen vapor keeps the bulb free of tungsten. Iodine reacts with the deposited tungsten to form the gas, tungsten iodide. The tungsten iodide gas then dissociates at the hot filament to redeposit the tungsten onto the filament surface and free the iodine for repeated operation. The tungsten-halogen lamp has been used as a source in a number of optical fiber sensors with or without interference filters.

3.3.2 Non-thermal sources

The non-thermal sources may be coherent or incoherent. For example, laser is a coherent source while light-emitting diode (LED) is an incoherent source. To understand the mechanism of light emission in these sources one needs to know the energy level diagram, photon transitions and the nature of transitions resulting in the monochromaticity *etc*. First we shall discuss the laser and then light-emitting diode.

Laser

When an electron moves from a higher energy level to a lower energy level in an atom, a photon is generated. The energy of the photon is given by

$$E = E_2 - E_1 = h\nu \tag{3.5}$$

where E_1 and E_2 are the energies corresponding to lower and higher energy levels.

In an atom, three kinds of transitions involving electromagnetic radiation are possible between two energy levels E_1 and E_2. If the atom is initially in the lower energy state E_1 then it can be raised to E_2 on exciting the electrons from lower energy level to the higher energy level by the absorption of photons of energy $h\nu = E_2 - E_1$ (see Fig. 3.2 (a)). This process is called induced or stimulated absorption. The atom does not remain in the excited state for a long time. The electron decays back to the ground state in about 10 ns and this happens spontaneously. The spontaneous transition from higher energy state E_2 to the lower energy state E_1 often leads to spontaneous emission of a photon of exactly the same energy $h\nu$ as the photons that excited the electrons in the first place as shown in Fig. 3.2(b). The process is called spontaneous emission. The light produced in this way radiates from the atoms in random directions. In the third possibility, an incident photon of energy equal to the energy difference between the two states ($E_2 - E_1$) interacts with the atom in the higher energy state and causes a transition from E_2 to E_1 state which results in the generation of a second photon of same energy. As a result, not only the two photons have same energy, frequency and wavelength, they are headed in the same direction and have the same polarization and phase. The third process is called induced or stimulated emission (Fig. 3.2(c)). Since the radiated light waves are exactly in phase with the incident ones so the result is an enhanced beam of coherent light. The spontaneous emission process is a random process and hence the light emitted from a large number of atoms using this process is incoherent because the radiated light waves are not in phase.

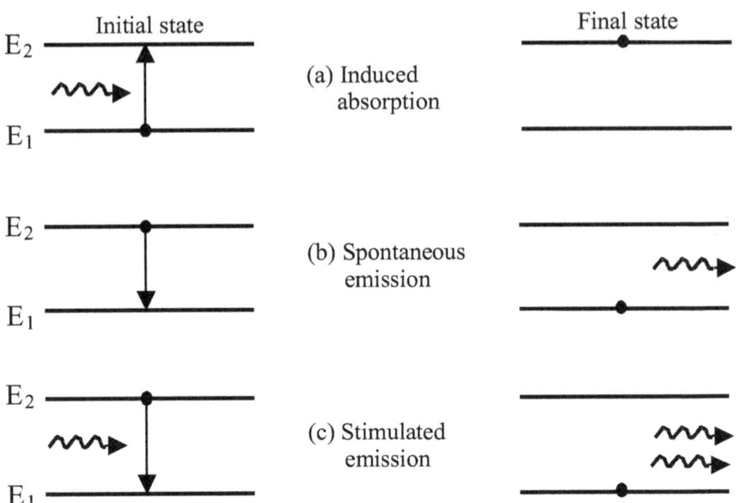

Figure 3.2: Different transitions between two energy levels in an atom. The filled circle indicates the state of the atom before and after transition.

In 1917, Einstein showed that induced emission has the same probability as induced absorption. That is, a photon of energy hν incident on an atom in the higher state E_2 has the same likelihood of causing the emission of another photon of energy hν as its likelihood of being absorbed if it is incident on an atom in the lower energy state E_1. Now consider an assembly of atoms having metastable state of excitation energy hν. The metastable state is that state in which an atom can linger for microseconds or even for milliseconds before decaying to a lower state. Suppose a majority of atoms are at the metastable state, a phenomenon called population inversion. If the light of frequency ν falls on the assembly of these atoms then there will be more induced emission from the metastable level than induced absorption by the lower level. The result will be an amplification of the incident light. This is the principle of a laser. There are a number of ways to produce a population inversion. It can be done by optical pumping or by the collisions of atoms. Figure 3.3 shows a three level laser system. The atoms in the ground state E_1 are pumped to state E_3 by photons of energy $h\nu_p = E_3 - E_1$. The atoms then decay spontaneously to the metastable state E_2 where they spend an unusually long time. It is from this metastable state that the stimulated emission or lasing takes place.

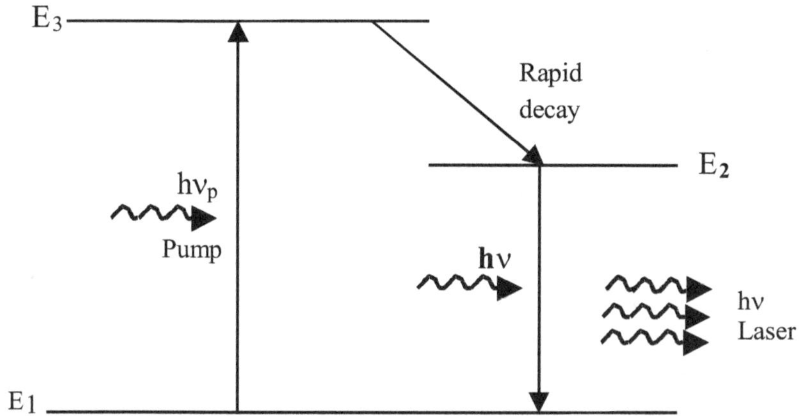

Figure 3.3: The three level laser system.

Gas laser

There is a wide range of gas lasers which could be used in optical sensing. The most frequently used are the helium-neon (He-Ne) laser, the argon and krypton ion lasers, to some extent the CO_2 laser and the excimer lasers. These lasers provide a wide variety of wavelengths in the region from the ultraviolet through the visible to the infrared. The most commonly used gas laser, at least in the laboratory, is the He-Ne laser. It is available at moderate price in a rugged configuration and provides good beam quality. Typical power outputs from He-Ne laser lie between 1 to 5 mW in a continuous beam. Because of its wide spread use in teaching laboratories and sensors, we shall discuss here only He-Ne laser.

The He-Ne laser Fig. 3.4 consists of a long and narrow discharge tube of diameter between 2 to 8 mm and length varying from 10 to 100 cm. The tube is filled with a mixture of about ten parts of helium to one part of neon. The neon atoms provide the lasing while the helium atoms provide an efficient excitation mechanism for the neon atoms. When an electrical discharge created by applying a high voltage (about 2 to 4 kV) is passed through the gas, the electrons which are accelerated down the tube collide with the helium and neon atoms and excite them to higher energy levels (see Figure 3.5). The helium atoms accumulate at F_2 and F_3 levels due to their long life times of about 10^{-4} and $5x10^{-6}$ sec respectively. Since the levels E_4 and E_6 of neon atoms have almost the same energy as F_2 and F_3, the excited helium atoms collide with the neon atoms in the ground state and excite the neon atoms to E_4 and E_6 levels. Since the pressure of helium is ten times that of neon, the levels E_4 and E_6 of neon are selectively populated as compared to other neon levels. The transition between E_6 and E_3 produces the very popular 632.8 nm line of the He-Ne laser. Neon atoms de-excite through the spontaneous emission from E_3 to E_2 ($\sim 10^{-8}$ sec). Since this time is shorter than the life time of level E_6 ($\sim 10^{-7}$ sec) one can achieve steady state population inversion between E_6 and E_3. The level E_2 is metastable and thus collects the atoms. The atoms from this level relax back to the ground level through the collisions with the walls of the tube.

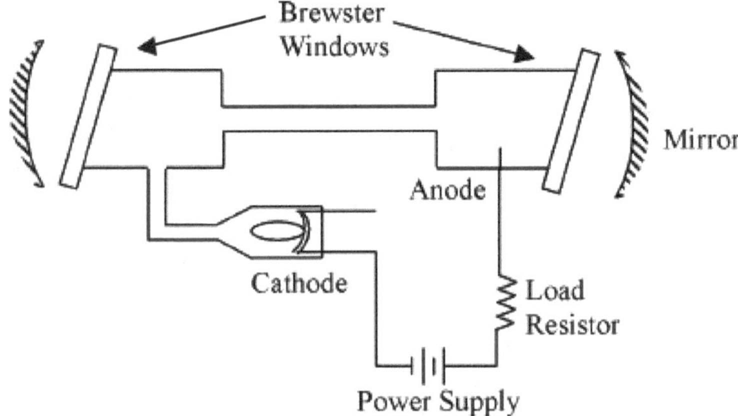

Figure 3.4: A typical He-Ne laser with external mirrors. The ends of the discharge tube are fitted with the Brewster windows. The load resistor limits the current once the discharge has been initiated.

The other two important wavelengths from the He-Ne laser are 1.15 μm and 3.39 μm, which correspond to the $E_4 \rightarrow E_3$ and $E_6 \rightarrow E_5$ transitions. All these transitions compete with each other and therefore one has to prevent the unwanted wavelengths from lasing. This can be achieved by using the multi-layer coated mirrors having wavelength dependent reflectance and forming resonant cavity.

If the resonant mirrors are placed outside the discharge tube then the reflections from the ends of the discharge tube can be avoided by placing the glass windows oriented at Brewster angle to the axis of the tube. In such a case the beam polarized in the plane of incidence suffers no reflection at the windows while the perpendicular polarization suffers reflection losses. This leads to a polarized output of the laser. The power output from the He-Ne laser is small (~ 1 mW to 5 mW) in a continuous beam. However, the radiation is extremely useful because it is highly collimated, coherent and has an extremely narrow line width. Unlike the light from conventional sources, such as tungsten halogen lamp, laser light is exceptionally monochromatic. A He-Ne laser is about 10 million times more monochromatic than the 589.6 nm emission line of a sodium vapor discharge lamp.

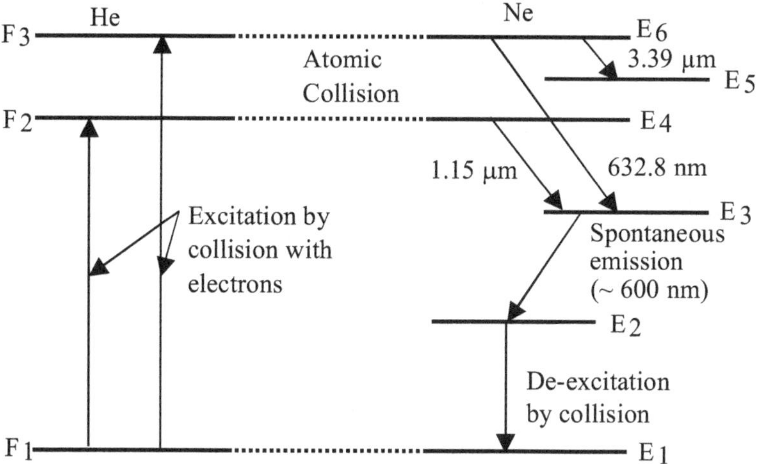

Figure 3.5: Energy levels of helium and neon taking part in the He-Ne laser.

The He-Ne laser described above illustrates one way of producing laser light. There are a number of other methods. However, the basic prerequisites for laser light are the presence of a metastable state, some mechanism of pumping atoms, a population inversion of atoms in the metastable state, stimulated emission and some kind of optical feedback to enhance the stimulated emission. Without these conditions, a laser would be just another discharge lamp.

Semiconductor laser (Laser diode)

If isolated atoms are brought together to form a solid then various kinds of interactions occur between neighboring atoms. The forces of attraction and repulsion between atoms find a balance at the proper interatomic spacing for the crystal. The energy levels of interacting atoms change into bands. The lower energy bands of the solid are occupied by electrons. The completely occupied or full energy bands are not important for determining the electrical properties of the solids. On the other hand, the electrons in the higher energy bands of the solid are important in determining its various physical properties. The two highest energy bands, called the valance and conduction bands, along with the forbidden energy gap (E_g) between them are important. The valence band might be completely filled, nearly filled or only half filled with electrons while the conduction band is never more than slightly filled. In some metals the highest two bands overlap to some extent resulting in somewhat different conducting properties (see Fig. 3.6 (a)). In insulators the valence band is completely filled with electrons while the conduction band is empty (see Fig. 3.6(b)). The conduction band is separated from the valence band by a large energy gap ($E_g \approx 4$ eV) and hence electrons cannot respond to an applied field and no conduction can occur. In intrinsic or pure semiconductors, the valence band is full at T=0 while the conduction band is empty. In other words, at T=0, all the states up to $E = E_F$ and none above E_F are occupied by electrons; E_F is called the Fermi energy. In the case of intrinsic semiconductor the Fermi energy level lies at the centre of the band gap and hence acts as a reference point. The band gap in semiconductors is small (~1 eV) and therefore at higher temperature some electrons in valence band get excited and move to the conduction band (see Fig.3.6(c)) leaving holes in the valence band. The electrons and holes allow conduction through the material and are called carriers. The width of the energy band gap is different in different semiconductors.

The electrical conductivity of an intrinsic semiconductor can be increased by adding doping element, or small percentage of an impurity element, to the semiconductor. The presence of the small traces of impurity elements can yield extra charge carriers which are free to move through the material. The impurity that creates more free electrons when doped in semiconductor is called donor impurity while the one that creates more holes is called acceptor impurity. When donor impurities are added, thermally excited electrons from the donor levels which are just below the conduction band (Fig. 3.7(a)) are raised into the conduction band to create an excess of electrons. This type of semiconductor is called n-type semiconductor and in these semiconductors the majority charge carriers are electrons. The excess of charge carriers in conduction band raises the Fermi level above the centre of the band gap. In the case of acceptor impurities, thermally excited electrons are raised from the valence band to the acceptor impurity levels leaving an excess of positive charge carriers *i.e.* holes in the valence band. Such type of semiconductor is called p-type semiconductor. In this case the majority carriers are holes and the Fermi energy level is below the centre of the band gap as shown in (Fig. 3.7(b)).

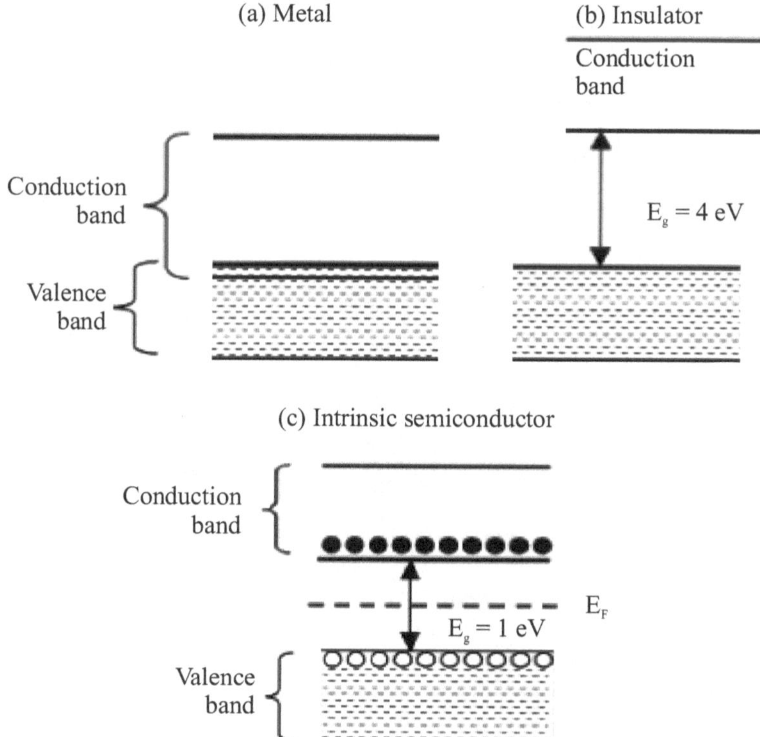

Figure 3.6: Energy bands in (a) metal, (b) insulator and (c) intrinsic semiconductor.

The intrinsic semiconductors are silicon, germanium, gallium arsenide, gallium phosphate, indium phosphate *etc.* The donor impurities are phosphorous, antimony, bismuth and arsenic. These have five valence electrons. The indium, tellurium, gallium have atoms with three valence electrons and can be used as acceptor impurities.

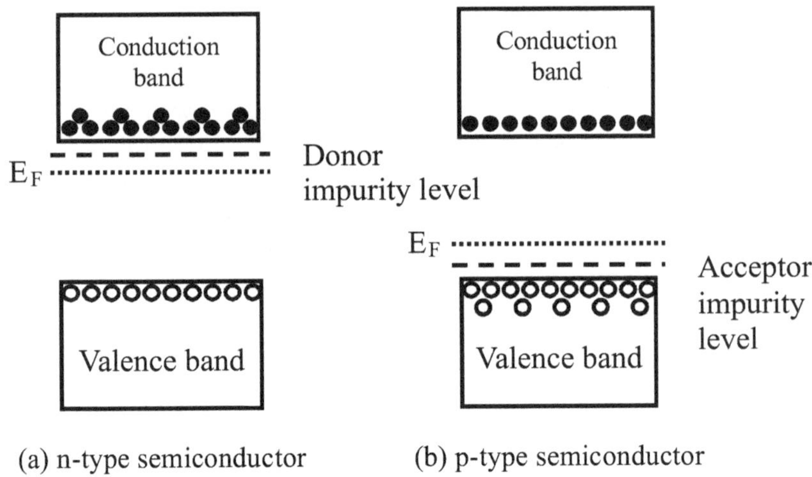

Figure 3.7: Energy band diagrams of (a) n-type (b) p-type semiconductors.

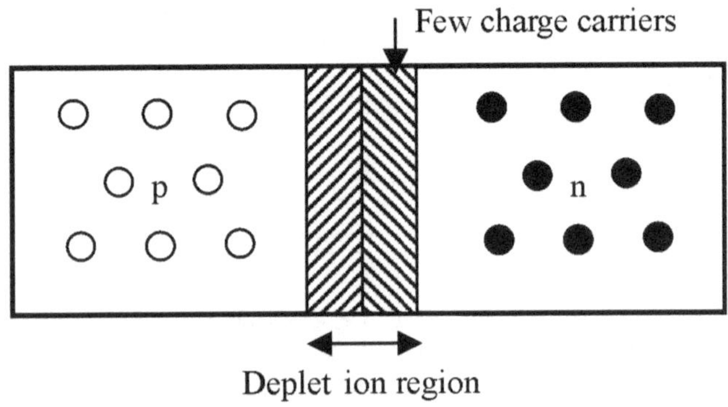

Figure 3.8: The p-n junction.

The combination of p-type and n-type semiconductors is used to design various kinds of semiconductor devices. The operation of most of the devices is based on the nature of junctions between p- and n-type materials. When a p-type material joins an n-type material, a depletion region occurs between them instead of a sharp interface as shown in Fig. 3.8. In this region electrons from the n-type material recombine with the holes of the p-type material so that very few charge carriers of either kind are present in this region. The Fermi energy is the same in both p- and n- regions. If it is not then electrons will flow to the region with vacant states of lower energy until E_F becomes equal. The width of the depletion region is typically ≈ 10^{-6} m. It depends on the carrier concentration (doping) in the p- and n-type regions and any external voltage applied across the junction. If an external voltage is applied with the p-end positive and the n-end negative (forward bias), the depletion region is reduced and electrons from n-type region and holes from p-type region flows more readily across the junction into the opposite type region (see Fig. 3.9). The holes and electrons meet in the vicinity of the p-n junction and recombine there. The recombination results in the release of energy approximately equal to the band gap energy.

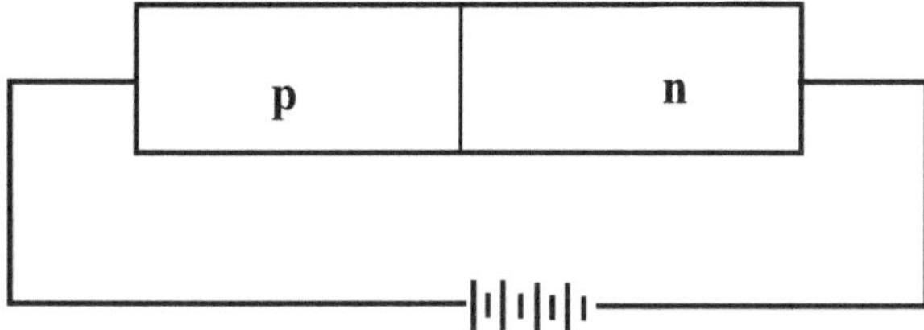

Figure 3.9: p-n junction with forward bias.

If the p-n junction is irradiated with light of frequency E_g/h then there can be following interaction processes: (a) an electron in the valence band may absorb it and gets excited to the conduction band and (b) a stimulated emission may occur in which the incident radiation stimulates an electron in the conduction band to make a transition to the valence band and in the process emit radiation of same frequency, E_g/h. If by some mechanism a large density of electrons is created in the bottom of the conduction band and simultaneously in the same region of space a large density of holes is created at the top of the valence band then an optical beam with a frequency slightly greater than E_g/h will cause a larger number of stimulated emissions as compared to absorption and thus can be amplified. In this case, the optical feedback is done by cleaving or polishing the ends of the p-n junction at right angles to the junction. When a current is passed through a p-n junction under forward bias the injected electrons and holes will increase the density of electrons in the conduction band and holes in the valence band. If the current is increased then at some value the stimulated emission rate will exceed the absorption rate and amplification will begin. As the current is further increased, at some threshold value, the amplification will overcome the losses in the cavity and the laser will begin to emit coherent radiation. This is how the laser diodes or semiconductor lasers emit light.

As mentioned above, to achieve stimulated emission, we need some initial radiation (external light source pump or some spontaneous emission) and a method for confining the generated photons to a small area to increase the chances of their collision with electrons and producing more transitions and so forth. In laser diode, the initial photons are generated by means of injected electrical current. The photons are confined to the active area by mirrors at the ends of the diode.

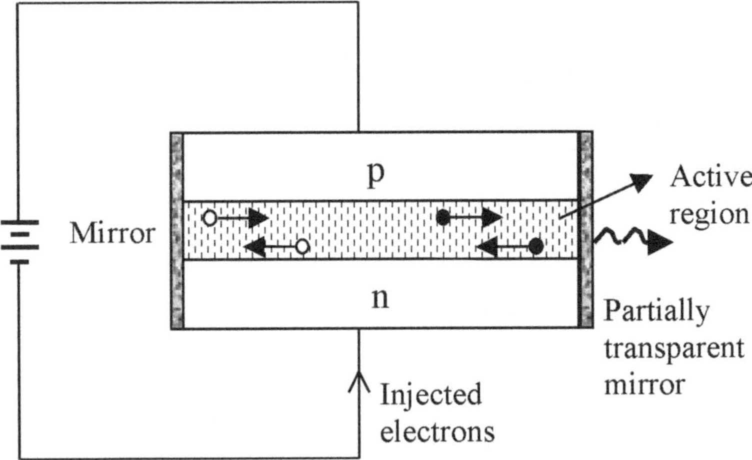

Figure 3.10: Basic semiconductor laser structure.

Figure 3.10 shows a basic semiconductor laser. The photons generated by the injected current travel to the edge mirrors and are reflected back into the active area. It causes the collision of photons and electrons and then produces more photons. These photons continue to bounce back and forth between the two edge mirrors. This process eventually increases the number of generated photons until lasing takes place. One of the edge mirrors is partially transparent to allow the generated light to leave the structure. The lasing takes place at particular wavelengths that are related to the length of the cavity, L. To be more precise, $L = n\lambda/2$, where n is an integer. This relationship will be satisfied for a number of wavelengths and hence a number of wavelengths will be emitted. When this happens the laser is called a multimode laser. If all, except one, peaks of the radiation are suppressed then the laser is called single mode laser.

The diode lasers are not 100% efficient because all of the electrical energy input does not go into creating photons. Some input energy goes into the semiconductor crystalline lattice. This elevates the bulk temperature of the semiconductor but does not contribute to lasing.

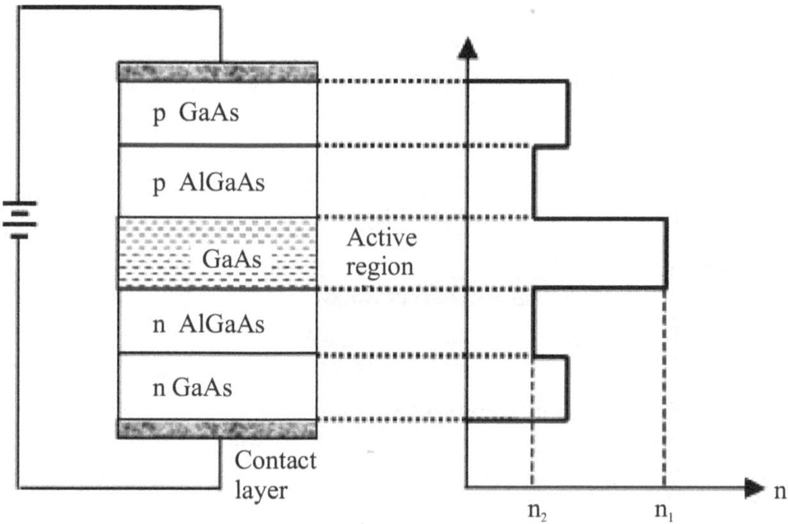

Figure 3.11: Laser diode with more than one p and n layers. On right side the refractive index profile is shown.

Modern laser diodes are formed of structures that contain several thin layers of varying composition. One example with more than one p and n layers is shown in Fig.3.11. The p and n type materials at the junction have light doping while the other two p and n materials have heavy doping. The two semiconductors used for p and n type of materials are GaAs and AlGaAs. GaAs has refractive index higher than AlGaAs. In addition to these, a thin layer of GaAs is also placed at the p-n junction. This layer acts as the active region. The refractive index profile across the layers is also shown in (Fig. 3.11). The refractive index of the active region (0.2-0.5 μm) is more than the surrounding p and n layers. These three layers form an optical waveguide and hence confines the light generated into the active region. The double layer restricts the electron concentration to the junction area and thus improves the radiation efficiency. In this laser diode the injected current flows through the full cross-section of the structure and hence the radiation emission occurs from the whole width of the active region.

The wavelength at which laser emits is strongly influenced by the temperature of operation. Further, the life time of the laser diode decreases exponentially with increasing temperature. Thus the laser diode must be in good thermal contact with a heat sink capable of dissipating the thermal load generated by the laser. The beam emitted from a semiconductor laser typically has an elliptically spatial profile. It is caused by diffraction.

Light emitting diode

The principle of light-emitting diode (LED) is different from that of the laser diode. As mentioned above, when a forward bias voltage is applied to a p-n junction the recombination of electrons-holes pairs occur which result in the release of energy. In silicon and germanium the recombination energy is absorbed by the material as a heat, but in certain other semiconductors a photon is emitted. The energy of the photon resulting from this recombination is equal to that associated with the band gap. The semiconductors in which light emission occurs are gallium arsenide (GaAs), gallium arsenide phosphor (GaAsP), gallium phosphor (GaP), aluminium gallium arsenide (AlGaAs) and indium gallium arsenide phosphor (InGaAsP). This is the principle of the light-emitting diode (LED). In LED, no amplification of light takes place and the light is produced by the process of spontaneous emission and therefore the light produced in LED is incoherent.

There are two basic types of light-emitting diodes that are commonly used for fiber optic sensors. These are surface-emitting diode and the edge-emitting diode. In surface emitting diode, light is emitted from the junction surface while in the case of edge-emitting diode the edge of the active area radiates. Figure 3.12 shows a simple light-emitting diode. It consists of one p and one n-type semiconductor. The p-region is relatively thin (about 1 μm). The recombination of electrons-holes occurs around the p-n junction. The light emitted from it radiates in all directions. The output light coming out from the thin p layer is very weak as compared to the total light generated at the junction. Thus this type of LED is very inefficient.

In the simple LED a single material and a single p and n interface has been used. To improve the efficiency of the LED, more than one semiconductor and p and n layers are used. The different

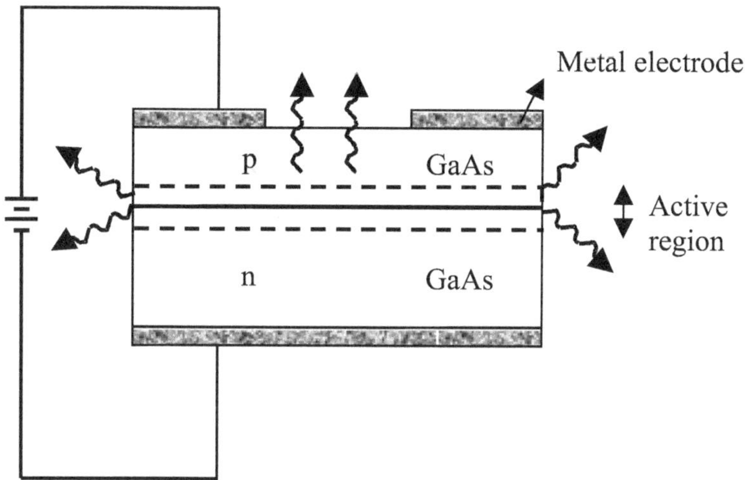

Figure 3.12: Simple light emitting diode

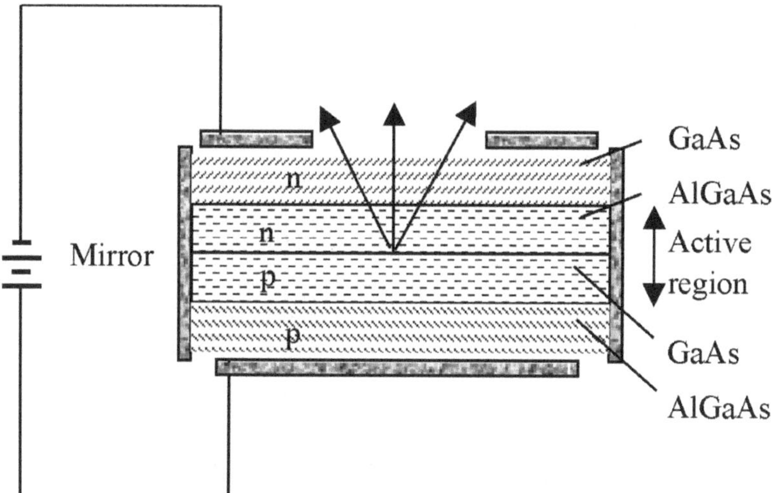

Figure 3.13: Surface-emitting LED.

semiconductors, for example AlGaAs and GaAs, have different refractive indices and hence help in confining the emitted light. Further the doping concentrations in the semiconductors are different. Different doping concentrations increase the concentration of holes and electrons in the active region. Fig. 3.13 shows a multi-layer surface emitting LED structure. Two different materials have been used for p and n regions. The lower p region serves as a reflector while the n-regions are transparent. In this case most of the power is radiated upward. A number of variations have been introduced on this structure to improve the radiation efficiency. Further to improve the coupling of light into an optical fiber the fiber is butted to the emitting surface. The surface emitting LED has a wide beam and therefore various methods are used to focus it onto the fiber end face. The edge-emitting LEDs have also been designed to produce a focused beam.

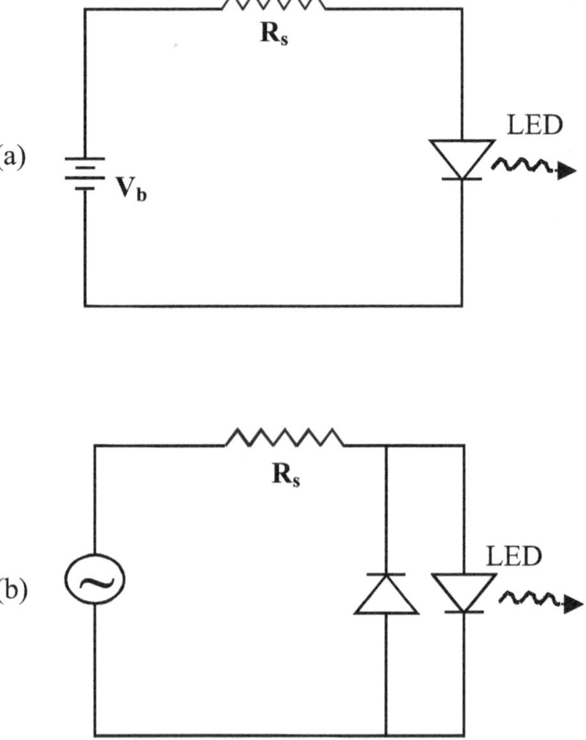

Figure 3.14: Simple LED drive circuits for (a) DC operation and (b) AC operation.

The electrical characteristics of LED's are essentially those of ordinary rectifying diodes. Typical operating currents are between 20 mA and 100 mA while the forward voltages vary from 1.2 V for GaAs to 2 V for GaP. The operating voltage is approximately equal to the built-in diode potential, which in turn is slightly less than the energy gap expressed in eV. Figures 3.14 (a) and (b) show the drive circuits for DC and AC voltage operation respectively. The current through the diode is limited by a series resistance R_S whose value may be calculated from

$$R_S = (V_b - V_d)/i_d \tag{3.6}$$

where V_b is the source voltage, V_d is the diode operating voltage and i_d is the desired diode current. In the AC circuit, a rectifying diode is placed across the LED to protect it against reverse bias breakdown.

Suggested Reading

1. A.K. Ghatak and K. Thyagarajan (1989) Optical electronics, Cambridge University Press.
2. J. Singh (1996) Optoelectronics: An introduction to materials and devices. McGraw-Hill Companies, Inc., Singapore.
3. J. Wilson and J.F.B. Hawkes (1989) Optoelectronics: An introduction, Prentice Hall.

4

Optical Detectors

4.1 Introduction

Apart from an optical source, the optical detector is another important component of the optical fiber sensor. The optical detector converts the optical signal to an electrical one. The signal is then processed either to record or display. There are various kinds of optical detectors but the commonly used in sensors are semiconductor diodes. In some of the cases photomultiplier tube has been used. In this chapter we shall describe the basic principles of optical detectors and their operational characteristics.

4.2 Basic requirements for detectors

To perform efficiently, an optical detector should have the following features:

Sensitivity: The detector should be highly sensitive. It should produce electrical current as large as possible in response to incident radiation at least in the operating wavelength region.

Response time: The response of the optical detector should be fast.

Linearity: The characteristics of the optical detector should be linear *i.e.* the incident energy and the generated electrical current should have linear relationship.

Noise: The signal to noise ratio should be as large as possible so that the smallest possible optical signal is detected.

Characteristics: The characteristics of the detector should be reliable, stable and insensitive to environmental conditions.

Cost: The detector should be cheap.

4.3 Classification of optical detectors

The optical detectors can be classified as thermal detectors and photon detectors.

4.3.1 Thermal detectors

In a thermal detector, incident light energy is converted into heat within the detector. The rise in temperature modulates the electrical signal. In these detectors, the energy of an individual photon has no relevance. It is only the total optical energy that is important so that the detector response is effectively independent of the wavelength. The examples of thermal detectors are thermoelectric detector, thermoresistive detector, pneumatic detector and pyroelectric detector.

4.3.2 Photon detectors

In a photon detector, an electrical charge carrier is liberated on the absorption of photon. For this the photon energy must be greater than some critical value. These detectors can further be classified into two classes such as photoemissive detectors and photoconductive detectors. Photoemissive detectors are not solid state. The examples of these detectors are photocathode, vacuum photodiode and photomultiplier tube. In these detectors the incident photon falls on a conducting surface called photocathode. The absorption of photon by this surface causes the emission of an electron which then travels through the free space and is captured by an anode. This produces a detectable current. The photoconductive detectors are solid state and are now most commonly used in sensors. These are also called photodiodes. We shall discuss here photomultiplier tube and photodiodes in more detail.

4.4 Photomultiplier tube

The photomultiplier tube is by far the most popular photoemissive detector in use today. This is because it is the most sensitive of optical detectors. A schematic diagram of a typical photomultiplier tube is shown in Fig. 4.1. It comprises a vacuum tube, a photoemissive surface (also called photocathode) and an electron multiplier. When a photon of sufficient energy strikes the photocathode, it dislodges an electron *via* the photoelectric effect. This electron is accelerated towards a series of electrodes (called dynodes) which are maintained at successively higher potentials with respect to the cathode (about 100 volts per stage). On striking a dynode surface each electron causes the emission of several secondary electrons which in turn are accelerated towards the next dynode and continue the multiplication process. If each dynode generates N secondary electrons per incident electron, and if there are M dynodes then the overall gain of the photomultiplier tube is N^M. If N=5 and M=9 then one obtains a gain of $2x10^6$. The multiplied photon induced electron current is finally collected by the anode where it can be readily detected. Gains of the order of 10^6 are the primary reason for their uniquely high sensitivity. The photocathode is mainly responsible for the operating characteristics of the photomultiplier tube. It consists of alkali metals with low work functions. The quantum efficiency of the photocathode depends on its material and the wavelength.

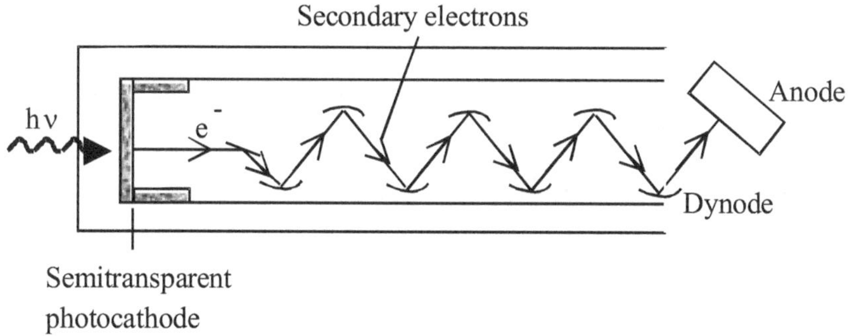

Figure 4.1: Schematic diagram of a photomultiplier tube.

4.4.1 Response time

The electrons emitted from dynodes take different times to reach anode. This is because electrons have spread in velocities when they are ejected from the cathode and in addition they traverse slightly different paths. Thus the photomultiplier tube has a finite transit time. Because of this a very fast pulse may not be detected accurately. Transit time spread increases with the increase in the number of dynodes. It can be reduced by proper design of dynodes. These days, photomultiplier tubes with 30 ns transit time are available.

4.4.2 Noise

There is no electrical system which does not suffer from noise. In a photomultiplier tube there are following sources of noise:

Dark current: Even when no radiation is falling onto the photocathode surface, thermionic emission gives rise to a dark current which is the main source of noise. This can be reduced by reducing the temperature of the photomultiplier tube.

Shot noise: When a current flows in a circuit the arrival rate of electrons at any point fluctuates slightly. This gives rise to fluctuations in the current flow at that point. The root mean square variation in the current is the *shot noise*. The r.m.s. current fluctuations with frequencies between f and f+Δf is given by

$$\Delta i = (2ie\Delta f)^{1/2} \tag{4.1}$$

where i is the total current (dark + signal) flowing, Δf is the noise bandwidth (in hertz) and e = 1.6×10^{-19} coulomb. The minimum detectable signal power (in watts) in the presence of a thermionic dark current, i_T, is given by

$$W_{min} = \frac{(2i_T e\Delta f)^{1/2}}{R_\lambda} \tag{4.2}$$

where R_λ is the responsivity of the photomultiplier tube. It is given by

$$R_\lambda = \frac{\eta e\lambda}{hc} \tag{4.3}$$

where η is the quantum efficiency of the cathode.

Multiplication noise: The noise at the anode is always greater than the shot noise. This is because of the statistical spread in the secondary electron emission coefficient about the mean value (N). Due to this the anode current increases by a factor $[N/(N-1)]^{1/2}$. If N increases this factor decreases.

Johnson noise: The Johnson noise arises because of thermal agitation of charge carriers within a conductor which results in the voltage fluctuation across the conductor. The r.m.s. value of this voltage fluctuation, ΔV, with frequencies between f and f+Δf across a resistance R at temperature T is given by

$$\Delta V = (4kTR\Delta f)^{1/2} \tag{4.4}$$

where k is the Boltzmann constant. In a photomultiplier tube such a noise voltage will appear across the anode load resistor. In practice, Johnson noise is smaller than the dark current shot noise.

4.5 Semiconductor photodetectors

The main types of semiconductor photodetectors are photo-conductors, photodiodes, PIN photodiodes and avalanchc photodiodes. We shall describe them one by one.

4.5.1 Photoconductor

A photoconductor is the simplest type of photodetector. It consists of a thick film of about 50-100 μm thickness of cadmium sulfide (CdS), or lead sulfide (PbS), or lead selenide (PbSe) between two electrodes. In some cases, mixtures, such as CdS with CdSe, are common. When light falls on the film, its resistance drops from extremely high levels, typically many megaohms or even gigaohms ($10^9\ \Omega$), to quite low levels. In bright sunlight it drops to tens of ohms. If a voltage is applied across the film, illumination leads to an output current, which is typically amplified by a simple circuit (see Fig. 4.2). The change in resistance in the film is caused by the basic phenomenon - the photoelectric effect. In a semiconductor, the conduction band, where electrons are free to move about, is separated from the valence band by an energy gap. When a photon knocks an electron out of the valence band into the conduction band via the photoelectric effect, the conductivity of the semiconductor increases Fig. 4.2(a). More the light, the more the electrons go into the conduction band and the greater is the conductivity. Since the photon must be of sufficient energy to allow the electron to jump across the band gap, the band gap of the semiconductor sets a lower limit on the frequency of radiation to which the photoconductor is sensitive. This is feasible provided that the photon energy exceeds the band gap energy of the semiconductor. Thus the following condition must be satisfied by the incident photon

$$h\nu \geq E_g \quad \text{or} \quad \lambda_c \leq \frac{hc}{E_g}$$

where λ_c is the cut off or the maximum wavelength. In the case of germanium the energy gap is 0.67 eV. This gives $\lambda_c \leq 1.85$ μm. It means that germanium detectors can be used only with wavelengths shorter than 1.85 μm. In contrast to germanium, silicon has a band gap of about 1.12 eV and is usable up to 1.11 μm and hence mainly in visible region. Photoconductors are typically sensitive to light intensity in the range of microwatts to milliwatts per square millimeter, and they are most sensitive to the portion of the visible spectrum to which the eye is sensitive.

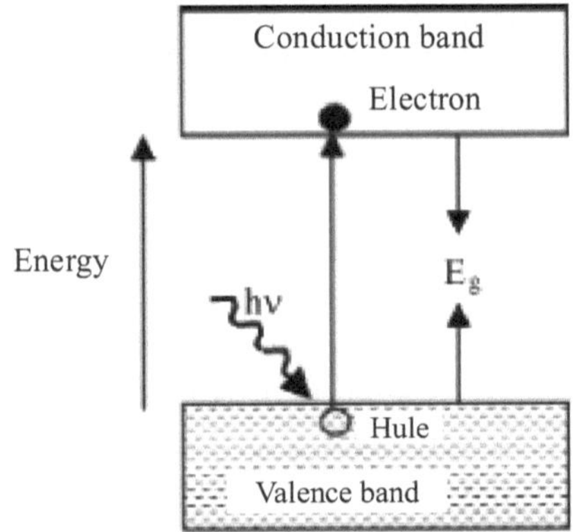

(a) Production of electron-hole

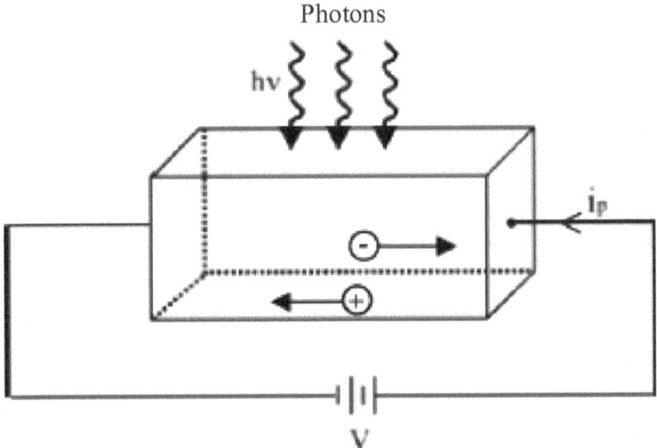

(b) Application of electric field across the semiconductor

Figure 4.2: (a) In photoconductors, absorption of photons increases the conductivity of semiconductors when electrons gain enough energy to jump the band gap (E_g), thus creating holes in the valence band and electrons in the conduction band. (b) Holes and electrons move in opposite directions in the presence of an electric field.

The advantages of photoconductors are their low cost, simplicity, and ruggedness. These detectors also have number of shortcomings. Photoconductors are not highly precise detectors. They typically have a nonlinear response, and their response times are often long ranging from tens to hundreds of milliseconds. In addition, their response often varies widely from detector to detector. In spite of these disadvantages, their basic robustness makes them appropriate for many applications in which the presence or absence of light is to be sensed.

4.5.2 Photodiodes

To achieve high sensitivity and repeatability, in addition to faster response time, a photodiode is used. It can be viewed as the inverse of a light-emitting diode (LED) described in chapter 3. In the case of photodiode the input is optical and the output is electrical. It is operated in reverse bias mode with the depletion region exposed to the optical energy. In the reverse bias operation the depletion region gets wider. The photons incident on the depletion region produce electrons-holes pairs which can then be moved by the applied voltage across the junction as shown in Fig. 4.3. There are few disadvantages which restrict the performance of the photodiode. One of these is that some of the photons may be absorbed in the p and n materials outside the depletion region. Another disadvantage is that the corresponding free charges move by diffusion at a very low velocity and most of them recombine along the way contributing very little to the resulting photocurrent. Thus it is desirable to increase the depletion region as much as possible so that more photons can be absorbed and hence high photocurrent is generated. To accomplish high performance PIN photodiodes have been developed.

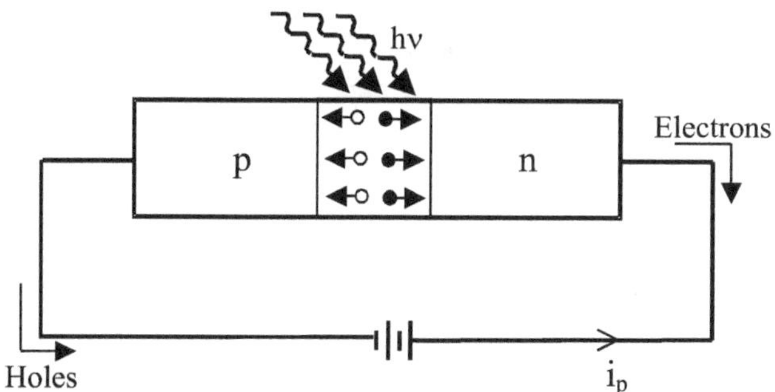

Figure 4.3: Operation of a photodiode.

4.5.3 PIN photodiodes

The silicon PIN photodiode is probably the most commonly used photodetector in sensors. It consists of three simple epitaxial layers in which a thick intrinsic semiconductor layer is sandwiched between a thin p-type semiconductor layer and a n-type layer - thus, the name PIN. Metallic layers are deposited on top and bottom of this semiconductor sandwich to provide electrical bias. The intrinsic material is silicon in the case of silicon photodiode and is germanium in the case of germanium photodiode. The intrinsic region has few free charges. It acts as a depletion region. A cross-section of a typical PIN photodiode structure is shown in Fig. 4.4. Light enters the photodiode through an antireflection coating. The refractive indices of the semiconductor materials are high, so that in the absence of a coating, the reflective losses would be substantial. The light incident first traverses a heavily doped low resistive p-layer before reaching the intrinsic semiconductor layer which is a high resistive region. The absorption of light occurs in the intrinsic semiconductor region which results in the production of electron-hole pairs. If the junction is reverse bias it accelerates the generated electrons and holes to the contact regions by the electric field. Any photons absorbed in p or n region will also produce electron-hole pairs but these will not make any significant contribution to the photocurrent because they are not swept along by the electric field and therefore will recombine.

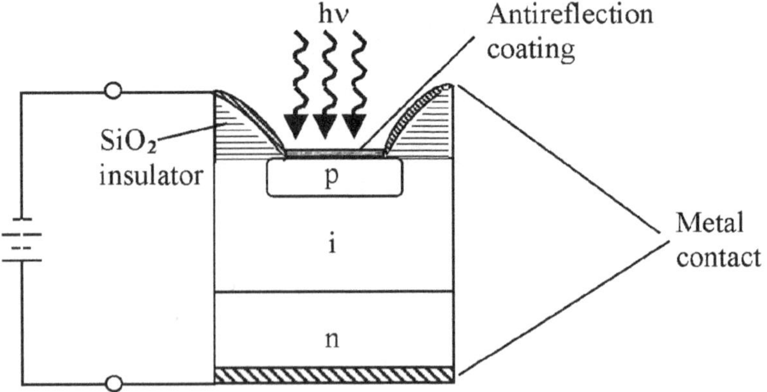

Figure 4.4: Schematic diagram of a PIN photodiode.

The depth to which a photon will penetrate a photodiode depends on its wavelength. Low wavelength photons traverse small distance as compared to long wavelength photons before absorption. Thus for short wavelength photons a thin layer of p-type material is used so that they can reach to the depletion region. For longer wavelength photons one needs a thick depletion region to capture the photons. For a given thickness of intrinsic region, the absorption efficiency decreases with wavelength. In the case of a silicon photodiode, the width of the intrinsic region ranges from 10 to 20 μm and the photodiode is highly efficient in the 0.8 to 0.9 μm wavelength region. At shorter wavelengths, the efficiency will fall because of absorption in the p-region. At longer wavelengths the light will pass through the intrinsic-region with rapidly decreasing absorption. Thus PIN photodiode can be optimized for a given wavelength. For longer wavelengths use, germanium may be used in place of silicon. Like photomultiplier tube, these photodiodes also suffer from dark current, shot noise and shot noise due to the incident light (photon-noise).

As discussed above, in the p-n junction photodiode the depletion region is small. By introducing intrinsic region, the depletion region was increased and the diode was named as PIN photodiode. The increase in depletion region increases the responsivity of the detector. Further, the response time reduces because there is no photogeneration of electrons in the p or n regions where charges move very slowly and secondly, since the depletion region is larger, the interelectrode capacity is smaller, leading to a faster response. The other advantage is low-voltage operation. The PIN photodiodes do not have an internal amplification mechanism. Their performance is influenced by the noise introduced by the external amplification circuitry which decreases its sensitivity.

4.5.4 Avalanche photodiodes

For high sensitivity and low light detection where photomultiplier tubes cannot be used a solid-state alternative - avalanche photodiodes (APDs) can be used. These devices produce, in a semiconductor, the multiplication of photoelectrons that PMTs produce in vacuum, but they have the advantage of miniaturization associated with semi-conductor technology. With their ability to count single photons and detect in short time intervals, APDs have also found applications in telecommunication systems. Their applications in optical fiber sensing are limited.

In an avalanche photodiode, a high reverse bias voltage is applied across the p-n junction. When exposed to light the optically generated carriers are accelerated to very high energies due to the applied field. These carriers excite additional carriers from the valence to the conduction band retaining enough energy to remain in the conduction band. The current produced by a single optical absorption is then multiplied through an 'avalanche' process very similar to that occurring in a PMT. The responsivity of these photodiodes lies in the range 25-100 A/W as compared to 0.5-0.8 A/W of the p-n photodiode. To produce the avalanche process APD requires a high reverse bias voltage between 40-400 V. With these high voltages, the response time is very short. The photodiode is highly sensitive to temperature fluctuations and bias voltages. For these reasons, they require both thermal stabilization and a very well regulated bias voltage for high-accuracy use.

We have seen various kinds of optical detectors which can be used for sensing applications. For most of the applications, silicon PIN photodiode is useful. If more sensitivity is needed then PMT or silicon APD can be used. It should be noted that no optical fiber sensor can give better performance than its photodetector. The photodetector and receiver noise set the resolution limit. The optical fiber sensors and photodetector technology are two diverse subjects. Therefore those who are active in R & D of optical fiber sensors, have to keep the track of developments taking place in detector technology in order to develop highly sensitive sensors.

Suggested Reading

1. J. Singh (1996) Optoelectronics: An introduction to materials and devices. McGraw-Hill Co., Inc., Singapore.
2. J. Wilson and J.F.B. Hawkes (1992) Optoelectronics: An introduction. Second edition. Prentice-Hall, New Delhi.

5

Evanescent Field Absorption Sensors

5.1 Introduction

Evanescent field absorption spectroscopy is a powerful well-established laboratory technique for chemical analysis. The technique utilizes the evanescent wave penetration at the boundary between two dielectric media in conditions of attenuated total reflection (ATR). One of the media, a thin slab-shaped non-absorbing crystal, is called a waveguide, while the other medium of lower refractive index is the absorbing sample being studied. At each reflection at the boundary of the crystal and the sample, the penetration of the evanescent wave of the guided ray into the absorbing sample gives rise to a reduction of the power propagating in the crystal. The degree of absorption depends on the amplitude of the evanescent field in the sample medium and the number of reflections within the waveguide. The former increases dramatically for incident angles approaching the crystal-sample critical angle while the latter increases with the decrease in the thickness of the waveguide. The number of reflections can also be increased by increasing the length of the waveguide. Thus, to achieve maximum absorption, an unclad optical fiber is well suited for evanescent wave absorption spectroscopy. This is because, unlike the crystal waveguide, the narrow core diameter of the fiber provides a large number of reflections per unit length. Further, the use of an optical fiber to transmit light to a section of its core in contact with an absorbing sample offers the possibility of remote sensing or in-line analysis. These advantages, together with the benefits of simplicity, reliability, flexibility and relatively low cost have motivated a number of research groups all over the world to investigate evanescent wave spectroscopy at the surface of an unclad optical fiber. This chapter will be devoted to the fiber optic sensors utilizing evanescent field absorption spectroscopy.

5.2 Evanescent wave

When the incident light with the angle of incidence greater than the critical angle hits the boundary between the optically dense and optically rarer media, the beam is totally reflected back into the dense medium. Interestingly enough, light is not instantaneously reflected when it reaches the interface. Rather, the superposition of the incident and reflected beams results in the formation of a standing electromagnetic wave (see Fig. 5.1).

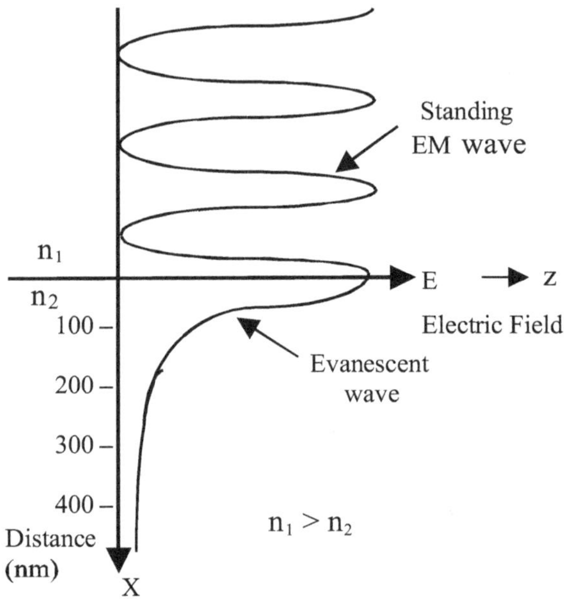

Figure 5.1: Evanescent wave at an interface of two media.

The maximum of the electric field amplitude is located at the interface but decays exponentially in the direction of an outward normal to the interface. The decaying field is called evanescent field. The electric field amplitude, E, at a distance x along the normal, is given by

$$E = E_o \exp(-x/d_p) \tag{5.1}$$

where E_o is the electric field at the interface and d_p is the penetration depth. The magnitude of the penetration depth is given by

$$d_p = \frac{\lambda}{2\pi n_1 \sqrt{\sin^2\theta - (n_2/n_1)^2}} \tag{5.2}$$

where λ is the wavelength of light in vacuum, θ is the angle of incidence with the normal to the interface, n_1 and n_2 are the refractive indices of the denser and rarer media, respectively. The penetration depth, d_p, is typically less than λ.

The evanescent wave propagates along the interface (*i.e.* along the z-axis) in the incident plane. This can be represented by several smaller light waves propagating parallel to the interface and having decreased intensity with increasing distance from the interface.

5.3 Evanescent wave for sensing applications

Figure 5.2 shows a plane wave incident upon the core-cladding interface. The direction of arrows indicates the flow of energy. The superposition of incident and reflected beams results in the formation of standing wave where the wavefront is perpendicular to the interface. The energy distribution in the core has its peak on the axis of the fiber. The energy in the cladding falls off exponentially. If a part of the cladding is removed from the fiber and is replaced by an absorbing fluid then a part of the evanescent field will be absorbed and hence the power transmitted by the fiber will decrease. This is the basis of the absorption sensors utilizing evanescent wave in an optical fiber.

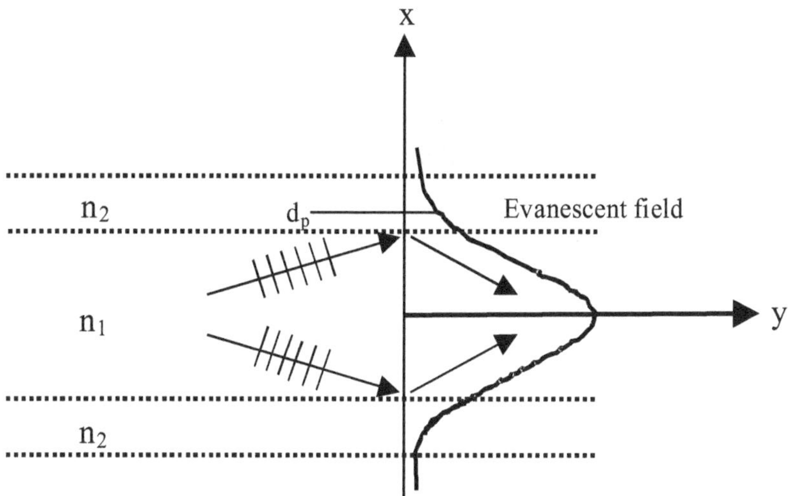

Figure 5.2: Evanescent wave phenomenon at the core-cladding interface in an optical fiber.

The absorption of evanescent field in an optical fiber has been exploited for the development of a large number of chemical and biological sensors. There are two approaches which have been adopted in these sensors. One, the evanescent wave can interact directly with the analyte if the wavelength of the light propagating coincides with its absorption band. Such sensors are called direct spectroscopic evanescent wave sensors. In the second approach, an intermediate reagent which responds optically (by fluorescence or absorption change) to the analyte, may be attached to the core of the fiber. These sensors are called reagent-mediated evanescent wave sensors. We shall discuss the second approach in the next chapter.

5.4 Evanescent absorption coefficient

Consider a step-index multimode optical fiber whose cladding has been replaced locally by an absorbing fluid (see Fig. 5.3). If P_o is the power transmitted by the fiber in the absence of an absorbing fluid then the power transmitted in its presence is given by

$$P = P_0 \exp(-\gamma L) \tag{5.3}$$

where L is the length of the unclad region and γ is the evanescent absorption coefficient of the fluid which can be written as

$$\gamma = NT \tag{5.4}$$

where N is the number of ray reflections per unit length of the fiber and T is the Fresnel transmission coefficient of the light at the interface of a loss-less core and lossy cladding. For the fiber of core radius ρ surrounded by the lossy cladding of refractive index n_2-ik, these parameters are given by

$$N = \cot\theta / 2\rho \tag{5.5}$$

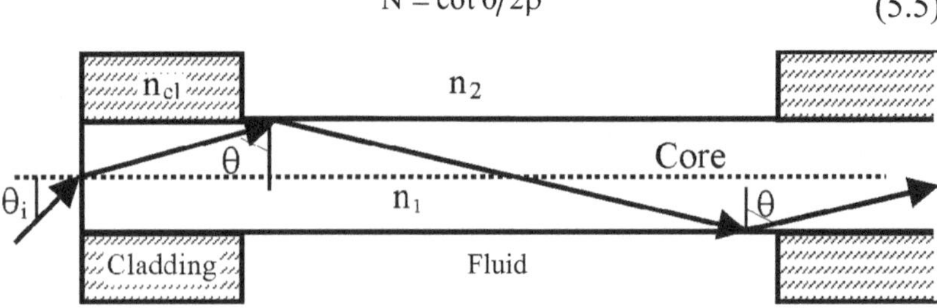

Figure 5.3: Propagation of a meridional ray in an optical fiber with uniform and straight core.

$$T = \frac{\alpha\lambda n_2 \cos\theta}{\pi\left(n_1^2 - n_2^2\right)\left(\sin^2\theta - \sin^2\theta_c\right)^{1/2}} \tag{5.6}$$

where $\alpha(= 4\pi k/\lambda)$ is the bulk absorption coefficient of the fluid at wavelength λ of the light source used. It is related to the concentration c of the absorbing fluid through the relation

$$\alpha = \varepsilon c \tag{5.7}$$

where ε is a constant and is the property of the substance dissolved in the fluid while the other symbols have their usual meanings. If θ_i is the angle of incidence of the ray at the end face of the fiber as shown in Fig. 5.3, then

$$\theta = \cos^{-1}\left(\sin\theta_i / n_1\right) \tag{5.8}$$

Substitution of Eqs. (5.5) and (5.6) in Eq. (5.4) gives

$$\gamma(\theta_i) = \frac{\alpha\lambda n_2 \sin^2\theta_i}{2\pi\rho\left(n_1^2 - n_2^2\right)\left(n_1^2 - \sin^2\theta_i\right)^{1/2}\left(n_1^2 - n_2^2 - \sin^2\theta_i\right)^{1/2}} \tag{5.9}$$

Equation (5.9) gives the value of evanescent absorption coefficient for different values of angle of incidence of the ray incident at the end face of the fiber. If all the guided rays are launched in the fiber then the effective evanescent absorption coefficient can be written as

$$\gamma_{eff} = \frac{\int_0^{\theta_{ic}} p(\theta_i)\gamma(\theta_i)d\theta_i}{\int_0^{\theta_{ic}} p(\theta_i)d\theta_i} \tag{5.10}$$

where

$$\theta_{ic} = \sin^{-1}\left[n_1 \cos\left(\sin^{-1}(n_{cl}/n_1)\right)\right]$$

is the critical (or the maximum) angle of incidence of the fiber at the end face, n_{cl} is the refractive index of the fiber cladding and $p(\theta_i)$ is the power distribution of the rays. If the light is launched into the fiber from a collimated source such as a laser using a microscope objective then

$$p(\theta_i) = \frac{\sin\theta_i}{\cos^3\theta_i} \tag{5.11}$$

5.5 Sensitivity

The evanescent wave absorption sensors are typically used to monitor the concentration of an absorbing fluid. In such sensors, the fractional change in output power per unit change in concentration of the fluid determines the sensitivity of the sensor. Mathematically, the sensitivity can be written as

$$S = -(1/P)(dP/dc) \tag{5.12}$$

where P is the power transmitted by the fiber as defined in Eq. (5.3). Substitution of Eq. (5.3) in (5.12) gives

$$S = L\,(d\gamma / dc) \tag{5.13}$$

Further, Eq. (5.9) can be written as

$$\gamma = \alpha\beta \tag{5.14}$$

where β is a quantity which depends on the design of the sensor and the refractive index of the fluid (n_2). Substitution of Eqs. (5.7) and (5.14) in Eq. (5.13) gives

$$S = L\varepsilon(\beta + c\, d\beta/dc) \tag{5.15}$$

If the refractive index of the fluid does not vary significantly with the concentration of the fluid within the required concentration range of the sensor then one can substitute $d\beta/dc = 0$ in Eq. (5.15) and in that case

$$S = L\varepsilon\beta = L\varepsilon(\gamma/\alpha) \tag{5.16}$$

Following conclusions can be drawn from Eq. (5.16):

1. For a fixed value of the bulk absorption coefficient or the concentration of the fluid, the sensitivity of the fiber optic evanescent field absorption sensor increases with the increase in the value of the evanescent absorption coefficient.
2. The sensitivity of the sensor increases with the increase in the length of the unclad fiber (*i.e.* the sensing region).

The fiber optic evanescent field absorption sensors have application in on-line fluid analysis. These sensors can be used to study the absorption spectra of liquids and pastes, to study the kinetics of a chemical reaction and also to monitor continuously the concentration of the fluid in a chemical plant. Further, these sensors can be used in the chemical industries for distributed sensing as well as for sensing at isolated locations (see chapter 13). Below we describe fiber optic evanescent field absorption sensors based on different kinds of probes.

5.6 Simple probe

Figure 5.4 shows the simplest evanescent field absorption sensor based on uniform and straight sensing probe[1]. Light from a source such as He-Ne laser is focused using a microscope objective on the input face of a plastic-clad silica (PCS) fiber. The numerical aperture (NA) of the objective is always chosen to be greater than the numerical aperture of the fiber if all the guided rays are to propagate in the fiber. The silicone cladding is removed from the middle portion of the fiber and the unclad portion of the fiber is kept in a glass cell having facilities for filling and draining of the absorbing fluid. The output end of the fiber is connected to a power meter. For the best performance of the sensor, the absorption peak of the fluid should be close to the peak wavelength of the emission spectrum of the light source. The transmitted powers are measured when the glass cell is filled with absorbing fluid (P) and the solvent (P_0) separately. These powers along with the length of the unclad portion of the fiber are used to calculate the evanescent absorption coefficient from Eq. (5.3). The sensor is calibrated by measuring the transmitted powers for different concentrations of the fluid.

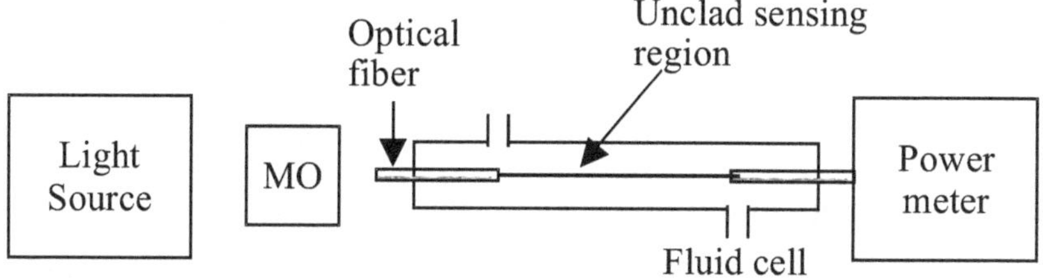

Figure 5.4: Experimental set up of a simplest fiber optic evanescent field absorption sensor[1]. The probe is straight and uniform in core radius.

Figure 5.5 shows the results obtained for the solution of cobalt chloride in isopropyl alcohol[2]. The experiments were done for the probes of different fiber core radii and numerical apertures. The variation is linear. In some of the cases (for example, methylene blue dye in water) these results are nonlinear. This is due to adsorption phenomenon[3]. If the curves are compared then one will find that for a fiber of fixed core radius the increase in numerical aperture increases the value of evanescent absorption coefficient and hence the sensitivity of the sensor. This occurs because as the numerical aperture of the fiber increases the refractive index of the fiber cladding decreases and it comes close to the critical angle of the sensing region. The closeness increases the penetration depth and hence the evanescent absorption coefficient. Figure 5.5 also shows that the decrease in the core radius increases the evanescent absorption coefficient. This is because of increase in the number of ray reflection in the unclad region in a fiber of smaller core radius.

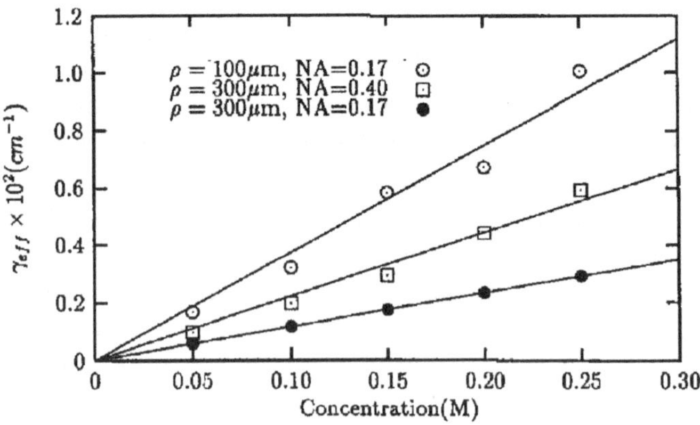

Figure 5.5: Variation of evanescent absorption coefficient with concentration for $CoCl_2$ solution. The experimental results are for three kinds of probes. (a) r = 100 μm, NA = 0.17, (b) r = 300 μm, NA = 0.40 and (c) r = 300 μm, NA = 0.17.

Reprinted from ref. 2 with kind permission of Springer Science and Business Media.

5.7 Tapered probe

The sensitivity of the fiber optic evanescent field absorption sensor based on straight and uniform probe and described above is very low. This is because of low penetration depth of the evanescent field. In terms of angle θ, the ratio (γ/α) (see Eq. 5.9) can be written as

$$\frac{\gamma(\theta)}{\alpha} = \frac{\lambda n_2 \cos\theta \cot\theta}{2\pi\rho\left(n_1^2 - n_2^2\right)\left(\sin^2\theta - \sin^2\theta_c\right)^{1/2}} \tag{5.17}$$

This equation implies that the ratio (γ/α) and hence the sensitivity can be enhanced by decreasing the core radius (ρ) as mentioned above. Further, if θ approaches θ_c, the sensitivity of the sensor increases. In the case of sensor described in Sec. 5.6, the angle of the guided ray remains constant throughout the sensing region. Therefore the minimum value that θ can have in the sensing region is equal to the critical angle of the fiber $\left(= \sin^{-1}\frac{n_{cl}}{n_1}\right)$. In practice $n_{cl} > n_2$, and hence θ cannot approach θ_c in the case of a uniform core probe. To achieve this a tapered probe, shown in (Fig. 5.6), is used in the fluid cell (Fig. 5.4). In the tapered probe, the radius of the core decreases along the direction of propagation thereby the number of reflections increases. Further, as the ray propagates through the taper its angle with the normal to the interface decreases and thus θ approaches closer to θ_c. Thus the sensitivity of the sensor is expected to increase. For a given fiber the increase depends on the taper ratio, R_T. In (Fig. 5.6), if θ_i represents the angle of the ray with the axis of the fiber then it gets transformed to an angle $\theta_a(z)$ inside the taper through the following relation[4,5]

$$\rho(z)\sin\theta_a(z) = \rho_i \sin\theta_i \tag{5.18}$$

where $\rho(z)$ is the taper radius at a distance z from the input end of the taper; ρ_i and ρ_o are the taper radii at the input and output ends, respectively. In terms of the angles with respect to the normal to the interface, Eq.(5.18) can be written as

$$\phi(z) = \cos^{-1}\left[\rho_i \cos\theta / \rho(z)\right] - \Omega \tag{5.19}$$

where $\phi(z)$ is the angle of the ray with the normal to the interface at a distance z from the input end of the taper; $\Omega = \tan^{-1}[(\rho_i - \rho_o)/L]$ is the taper angle; L is the length of the taper and

$$\rho(z) = \rho_i - \frac{z}{L}(\rho_i - \rho_o) \tag{5.20}$$

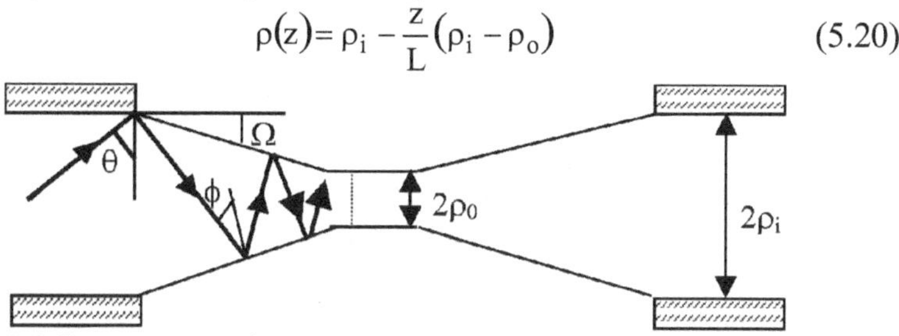

Figure 5.6: Propagation of a meridional ray in tapered region of an optical fiber.

The evanescent absorption coefficient in the case of taper, for the power distribution of the source given by Eq. (5.11), would be given by

$$\gamma_T(\theta_1,\theta_2,\rho_o)=\frac{\int_0^L dz\int_{\phi_1(z)}^{\phi_2(z)}\frac{\sin\theta\cos\theta}{\left(1-n_1^2\cos^2\theta\right)^2}\gamma(\theta,z)d\theta}{\int_0^L dz\int_{\phi_1(z)}^{\phi_2(z)}\frac{\sin\theta\cos\theta}{\left(1-n_1^2\cos^2\theta\right)^2}d\theta} \tag{5.21}$$

where

$$\phi_1(z)=\cos^{-1}\left[\rho_i\cos\theta_1/\rho(z)\right]-\Omega$$

and

$$\phi_2(z)=\cos^{-1}\left[\rho_i\cos\theta_2/\rho(z)\right]-\Omega$$

Substitution of Eq. (5.17) in Eq. (5.21) gives

$$\gamma_T(\theta_1,\theta_2,\rho_o)=\frac{\alpha\lambda n_1}{\pi\left(n_1^2-n_2^2\right)}\frac{\int_0^L\frac{dz}{\rho(z)}\int_{\phi_1(z)}^{\phi_2(z)}\frac{\cos^3\theta}{\left(1-n_1^2\cos^2\theta\right)^2\left(n_{12}^2\sin^2\theta-1\right)^{1/2}}d\theta}{\int_0^L\left[\frac{1}{1-n_1^2\cos^2\phi_1(z)}-\frac{1}{1-n_1^2\cos^2\phi_2(z)}\right]dz} \tag{5.22}$$

where $n_{12}=n_1/n_2$. It may be noted from Eq. (5.19) that $\phi(z)$ depends on $\rho(z)$ for a given value of θ. As $p(z)$ decreases $\phi(z)$ also decreases. A ray entering the taper region will remain guided if the following condition is satisfied:

$$\phi(z)\geq\sin^{-1}(n_2/n_1),\qquad 0\leq z\leq L \tag{5.23}$$

Thus, for given values of n_2, L and θ_1, there is a lower limit of of ρ_o.

To fabricate a tapered probe, the cladding is removed from the middle portion of the fiber. It is then firmly attached onto two metal plates which move in equal amount but in opposite directions with the help of an electrically driven motor. The unclad portion of the fiber is exposed to the flame and at the same time the plates on which the fiber is attached are allowed to move apart by equal amount through the controlling unit. The speed of the motor is carefully adjusted to obtain a uniform taper. The taper ratio R_T (= ρ_i/ρ_o) and the uniformity of the taper are checked by using a traveling microscope of high resolving power.

Figure 5.7 shows the results of the experiments carried out on the solution of cobalt chloride in isopropyl alcohol for different taper ratio of the probe. It can be seen that as the taper ratio increases the sensitivity increases[6]. The sensitivity of the tapered probe also depends on the taper profile[5]. It can further be increased if a combination of double taper is used[7].

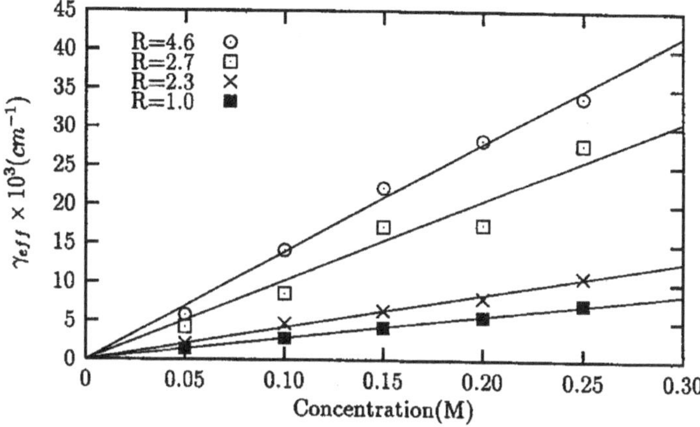

Figure 5.7: Variation of evanescent absorption coefficient with concentration of $CoCl_2$ solution for the sensor based on tapered probe. The results are for the fiber of core radius 300 μm and numerical aperture 0.40 for taper ratio 4.6 (o), 2.7 ( ), 2.3 (x) and 1.0 (▪). Reprinted with permission from ref. 6.

5.8 U-shaped probe

The sensitivity of the evanescent field absorption sensor can also be increased if the sensing region is bent as shown in Fig. 5.8. The bending reduces the angle of the ray with the normal to the interface in the bent region. Thus the angle θ in the fiber approaches close to the critical angle of the unclad bent sensing region. To obtain the evanescent absorption coefficient in the case of U-shaped probe, the two dimensional approach is considered[8]. Under this approach, the rays in the sensing region will be mainly meridional and will confine in the plane of bending. Since the skew rays are not considered, the value of the evanescent absorption coefficient will be more than the actual case. In the case of U-shaped probe there can be two possibilities: either a ray can undergo total internal reflection both at the outer and inner surfaces (continuous ray) or it undergoes total internal reflection at the outer surface only and does not reach the inner surface at all (broken ray). The angle of the ray at the outer surface of the bent core is given by

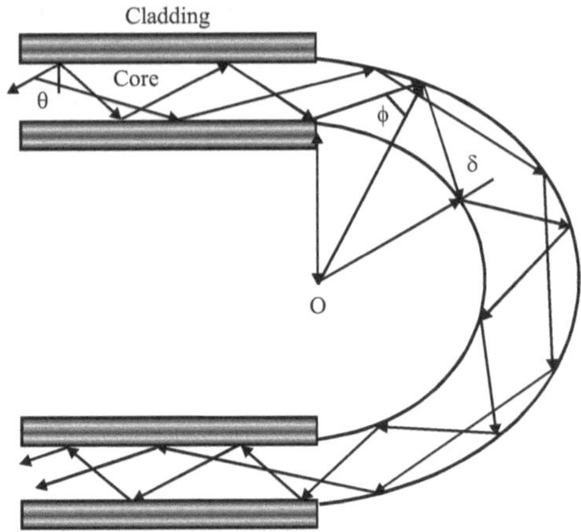

Figure 5.8: Propagation of meridional rays in an optical fiber with U-shaped probe.

$$\sin\phi = \left(\frac{R_B + h}{R_B + 2\rho}\right)\sin\theta \tag{5.24}$$

where h is the distance at which the ray is incident at the input face of the U-shaped region from the core-cladding interface and R_B is the bending radius of the probe. Similarly, the angle δ of the ray at the inner surface of the bent region is given by

$$\sin\delta = \left(\frac{R_B + h}{R_B}\right)\sin\theta \tag{5.25}$$

The effective evanescent absorption coefficient for the outer surface of the U-shaped probe can be written as

$$\gamma_{eff}(\phi_1, \phi_2)_{outer} = \frac{\alpha\lambda n_1}{4\pi\rho\left(n_1^2 - n_2^2\right)} \frac{\int_0^{2\rho}\int_{\phi_1}^{\phi_2} \frac{\cos^3\theta d\theta dh}{\left(1 - n_1^2\cos^2\theta\right)^2\left(n_{12}^2\sin^2\theta - 1\right)^{1/2}}}{\int_0^{2\rho}\int_{\phi_1}^{\phi_2} \frac{\sin\theta\cos\theta d\theta dh}{\left(1 - n_1^2\cos^2\theta\right)^2}} \tag{5.26}$$

where

$$\phi_1 = \sin^{-1}\left[\frac{(R_B + h)n_{cl}}{(R_B + 2\rho)n_1}\right]$$

and

$$\phi_2 = \sin^{-1}\left(\frac{R_B + h}{R_B + 2\rho}\right)$$

Similarly, for the inner surface of the U-shaped probe, the effective evanescent absorption coefficient can be written as

$$\gamma_{eff}(\delta_1, \delta_2)_{inner} = \frac{\alpha\lambda n_1}{4\pi\rho\left(n_1^2 - n_2^2\right)} \frac{\int_0^{2\rho}\int_{\delta_1}^{\delta_2} \frac{\cos^3\theta d\theta dh}{\left(1 - n_1^2\cos^2\theta\right)^2\left(n_{12}^2\sin^2\theta - 1\right)^{1/2}}}{\int_0^{2\rho}\int_{\delta_1}^{\delta_2} \frac{\sin\theta\cos\theta d\theta dh}{\left(1 - n_1^2\cos^2\theta\right)^2}} \quad (5.27)$$

where

$$\delta_1 = \sin^{-1}\left[\frac{(R_B + h)n_{cl}}{R_B n_1}\right]$$

and $\delta_2 = 90^o$. The effective evanescent absorption coefficient in the case of U-shaped probe is given by

$$\gamma_{eff} = \gamma_{eff}(\phi_1, \phi_2)_{outer} + \gamma_{eff}(\delta_1, \delta_2)_{inner} \quad (5.28)$$

If the total internal reflection occurs only from the outer surface then the second term in above equation vanishes. The values of γ_{eff}/α for different bending radii of the probe have been calculated[8]. The following values of the various parameters were used: $n_1 = 1.457$, $n_2 = 1.335$, $\rho = 100$ μm and $\lambda = 663$ nm. The calculated values for two different values of the numerical aperture of the fiber are tabulated in Table 5.1.

Table 5.1: Calculated values of evanescent absorption coefficient for U-shaped probe

Bending radius	γ_{eff}/α (NA = 0.2)	γ_{eff}/α (NA = 0.4)
∞ (straight probe)	$1.06\text{x}10^{-4}$	$5.25\text{x}10^{-4}$
1.5 cm	$2.54\text{x}10^{-4}$	$7.04\text{x}10^{-4}$
1.0 cm	$3.42\text{x}10^{-4}$	$8.35\text{x}10^{-4}$
0.5 cm	$6.84\text{x}10^{-4}$	$14.68\text{x}10^{-4}$

Following conclusions can be drawn from the results tabulated in Table 5.1:

1. As the bending radius decreases the value of the ratio γ_{eff}/α increases. This implies that the sensitivity of the probe increases with the decrease in the bending radius.

2. As the numerical aperture of the fiber increases the value of the ratio γ_{eff}/α increases. This implies that the sensitivity of the probe increases with the increase in the numerical aperture of the fiber.

These results were also verified experimentally[2]. To fabricate a U-shaped probe for the experiment, first the cladding is removed from a small portion of the central region of the multimode PCS fiber. The unclad part is then exposed to a flame and slowly bent till it becomes U-shaped. The temperature of the flame to be used depends on the size of the fiber core. For 600 μm core fiber, the temperature of the flame is kept around 700°C while for 200 μm core fiber, the temperature of the flame is kept in the range 400 to 500°C. The temperature of the flame is controlled by the proper mixing of LPG and O_2 gases. The uniformity of the core diameter and the bending radius of the probe are checked by using a traveling microscope. Figure 5.9 shows the opto-electronic system constituting the U-shaped sensor[2,8,9]. The measurement procedure is the same as discussed above for uniform and tapered probes. The light from a He-Ne laser is focused using a microscope objective on the input face of the fiber. The U-shaped probe is fixed in the cover of a glass cell using an adhesive paste. Fig. 5.10 shows the results of the experiments carried out on the solution of cobalt chloride in isopropyl alcohol for different bending radii of the U-shaped probe[9].

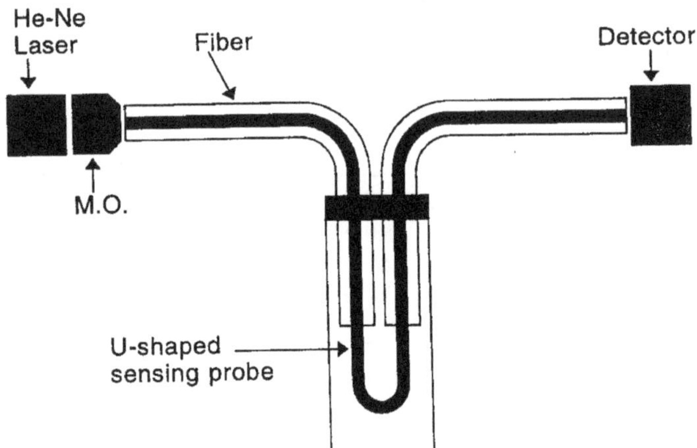

Figure 5.9: Experimental set up of a fiber optic evanescent field absorption sensor based on U-shaped probe. Reprinted from ref. 2 with kind permission of Springer Science and Business Media.

It can be seen that as the bending radius decreases the sensitivity increases. However the increase occurs up to a certain value of bending radius and after that it decreases. The upper limit on the sensitivity occurs due to the cross coupling of power between input and output arms of the U-shaped probe. The optimum bending radius depends on the numerical aperture and the core radius of the fiber. For a given bending radius, the sensitivity is higher for the fiber of higher numerical aperture. Further, for a given probe, the sensitivity increases if the refractive index of the solvent is increased[8].

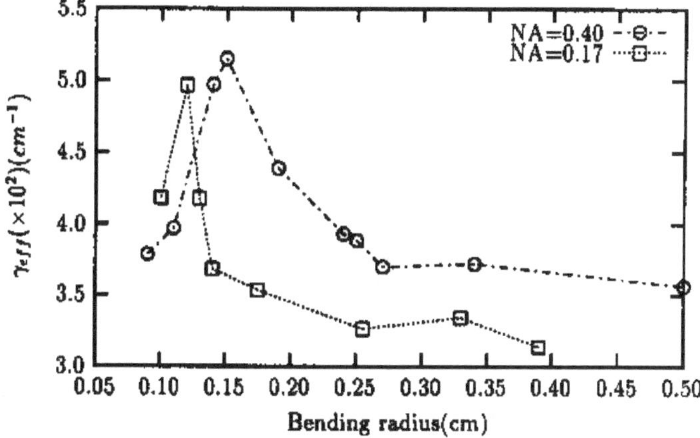

Figure : 5.10. Variation of evanescent absorption coefficient with bending radius of the U-shaped probe. The fiber is of 600 μm core diameter and the concentration of $CoCl_2$ is 0.20 M. The results are for fibers of two different numerical apertures. Reprinted from ref. 9 with permission from Elsevier.

5.9 Selective ray launching

The sensitivity of the evanescent field absorption sensor can further be increased if a selected ray is launched in the fiber[1]. In this case, γ/α is given by Eq. (5.17). The fiber optic sensor that uses selected rays is shown in Fig. 5.11. The light is launched into the fiber directly from a collimated source such as the He-Ne laser. To launch rays with different angles, the glass cell is kept on a rotation stage such that the input end of the fiber is positioned on the axis of the rotation stage. Rest of the arrangement is the same as in Fig. 5.4. The sensitivity increases many fold by using selected ray in the uniform and straight core fiber probe. In the case of tapered probe with θ_i=20° and R_T = 1.67, the sensitivity of the sensor increases by 7-times of the sensor with uniform core and all the guided rays launched. The increase occurs because all the rays are not guided in the fiber and the sensitivity depends on the angle of the ray with the normal to the core-cladding interface. The sensitivity is maximum for the ray that makes the angle close to the critical angle of the sensing region. The rays making angles higher than the critical angle in the sensing region can be blocked using an annular beam mask before launching in the fiber[10]. In this case there will not be any need of rotation stage.

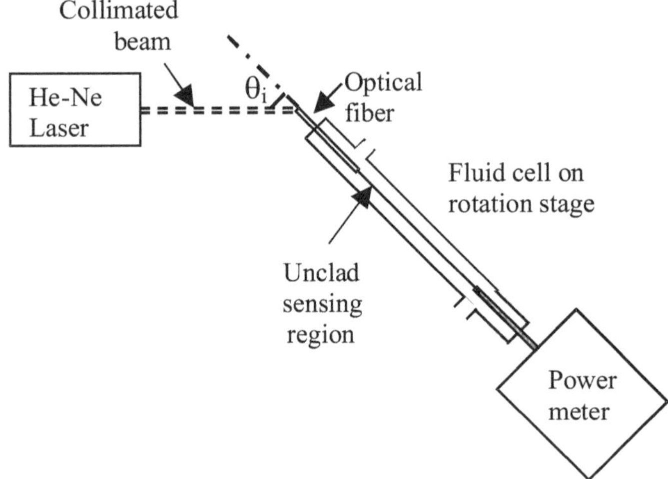

Figure 5.11: Fiber optic evanescent field absorption sensor based on selected ray launching[1].

5.10 Side polished half block

In the experiments on the sensors described above multimode fibers are used. However single mode fibers can also be used for the evanescent field absorption sensor. Figure 5.12 shows an optical fiber sensor based on the use of side polished optical fiber. The sensing region is a section of a single mode fiber, the cladding of which has been partially removed by polishing. The method is described in Sec. 6.5. The polishing technique was initially developed for the fabrication of monomode couplers. In this technique, the fiber is precisely located in an anisotropically etched silicon wafer or replica or quartz block and is then longitudinally polished. In consequence, the core of the bent fiber is brought near the surface, in the middle of the device. This portion provides access to the evanescent field on mechanically stable device.

The sample fluid is kept on the polished region of the fiber and the measurements are carried out similar to other probes described earlier. The main advantage of this sensor is that a very small amount of sample is needed because of very tiny probe.

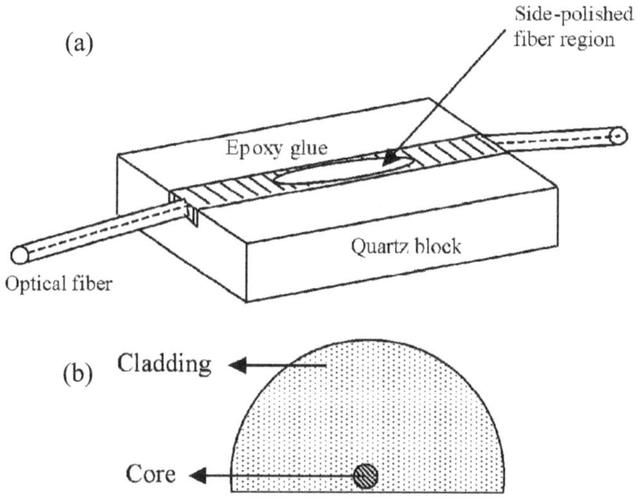

Figure 5.12: Side polished fiber block for evanescent field absorption sensor. (a) Global view and (b) Transverse view.

5.11 Evanescent field sensors with absorption in infrared region

Plastic clad silica (PCS) fibers used above for evanescent field absorption sensor cannot be used to sense the analytes that exhibit their fundamental absorption in the mid infrared region. This is because the transmission window of the silica fiber is limited only up to ~3 μm to infrared side while most of the chemical substances and gases have strong fundamental vibration absorption beyond this limit. Thus for this kind of applications, fibers of those materials which are transparent in mid infrared region are needed. Recently infrared transparent fibers have been developed for telecommunication and CO_2 laser power transmission purposes. The most competitive fibers are fluoride glass, chalcogenide glass and certain crystalline materials such as alkali halides. The chalcogenide fibers are transparent in the spectral range 2 to 11 μm while the silver halide fibers are transparent in the spectral range 2 to 20 μm.

An evanescent field absorption sensor that uses chalcogenide fiber was developed[11] for sensing acetone, ethyl alcohol and sulfuric acid in water. These chemicals are important industrial chemicals but pose certain environmental threats. Their fundamental vibrations are located in the 6-12 μm wavelength region and therefore chalcogenide glass fiber was used. Further, contrary to other infrared materials, chalcogenide glasses show good durability against water, various acids and some solvents. Due to these merits, it can be immersed directly into the aqueous solutions of acetone, ethyl alcohol and sulfuric acid without being damaged. In the experiment an unclad chalcogenide fiber with 380 μm core diameter is combined with a Fourier transform infrared (FTIR) spectrometer to monitor the concentration of these chemicals. Combining optical fiber to FTIR spectrometer offers a new dimension in flexibility which efficiently brings the spectrometer to the sample rather than requiring the sample to be removed and presented to the spectrometer. In addition, the spectra at various stages of the chemical processes can be measured using FTIR spectrometer. Further, the efficiency of transferring energy into the fiber with an FTIR spectrometer is large. Use of optical fiber increases the length of the sample cell which is not possible in the spectrometer. The middle and hence the sensing part of the fiber is encapsulated by the glass sample cell. The infrared beam of the FTIR spectrometer is focused onto the endface of the chalcogenide fiber. The output signal at the other end of the fiber is detected by a HgCdTe (mercury cadmium telluride), an infrared detector. Solutions containing various amounts of analytes in deionised water were prepared for the measurement and calibration. The first measurement is recorded with an empty sample cell to eliminate the effect of the fiber and atmospheric absorptions. The strongest absorption peaks of acetone, ethyl alcohol and sulfuric acid occur at 1710 (5.8 μm), 1045 (9.6 μm) and 1150 (8.7 μm) cm^{-1}, respectively. To calibrate the sensor for a given species, a corresponding wavelength is used and the output signal is recorded for different concentrations of the species. A linear relationship is observed between the evanescent absorption and the concentrations of various analytes. The minimum detection limit for acetone, ethyl alcohol and sulfuric acid are 5, 3 and 2 vol.% respectively with a sensor length of 15 cm. The presence of water affects the detection capacity in the case of acetone-water mixture because the absorption peak of acetone is close to that of water (1650 cm^{-1}).

It may be noted that the wavelength of light used for the above study ranges from 6 to 10 μm. These wavelengths are much larger than the visible light. According to Eq. (5.2), increase in wavelength increases the penetration depth and hence the sensitivity of the sensor. However, the refractive index of chalcogenide fiber used for this study is 2.972 at 10.6 μm. This is much larger than the refractive index of acetone (1.357), ethyl alcohol (1.359) and sulfuric acid (1.437). As described earlier, if the refractive index of the surrounding approaches that of the fiber core, the penetration depth increases. Here the difference in two refractive indices is large as compared to silica fiber and hence it will decrease the penetration depth. Thus the advantage of having large wavelength is lost by the high refractive index of chalcogenide. The evanescent field absorption spectroscopy has also been used for the detection of various gaseous species such as hexane, trichloro-trifluoro-ethane, methane, acetone as well as their mixtures[12]. The absorption peaks of these gases lie in the infrared region. The chalcogenide fiber coupled to FTIR spectrometer was used for their sensing.

5.12 Porous cladding

If the analyte is a gas then the sensitivity of the evanescent field absorption sensor is very low because the refractive index of gas is much lower than that of the fiber core. This causes increased confinement of the guided mode and significantly reduces its evanescent field in the unclad region and hence the sensitivity. The sensitivity of the sensor cannot be increased much by tapering or bending the probe. For achieving maximum sensitivity the tapering ratio has to be very large which is difficult to achieve and the probe will be highly fragile. Further, for the second kind of probe, the bending radius should be very small. This is limited by the actual fiber radius. The problem can be solved if a porous layer of glass or polymer whose effective refractive index is close to but less than that of the core, is deposited on the fiber core (see Fig. 5.13). The deposited layer will act as the cladding. Since the refractive index of the cladding is close to that of the core, the evanescent field will be more in the cladding and because of being porous the analyte (*i.e.* the gas) can diffuse in the cladding through pores and interacts with the evanescent field present. Thus the sensitivity of the sensor can increase.

The diffusion of the gas in porous cladding and the increase in evanescent absorbance has been studied mathematically[13]. To study this, a multimode optical fiber with porous cladding was considered. Let a and b be the radii of the fiber core and porous cladding, respectively. If light from a monochromatic source is launched into the fiber then the electric field E at a distance r from the fiber axis is related to the core field E_o through the following relation

$$E(r) = E_o \exp\left[-\frac{(r-a)}{d_p}\right] \tag{5.29}$$

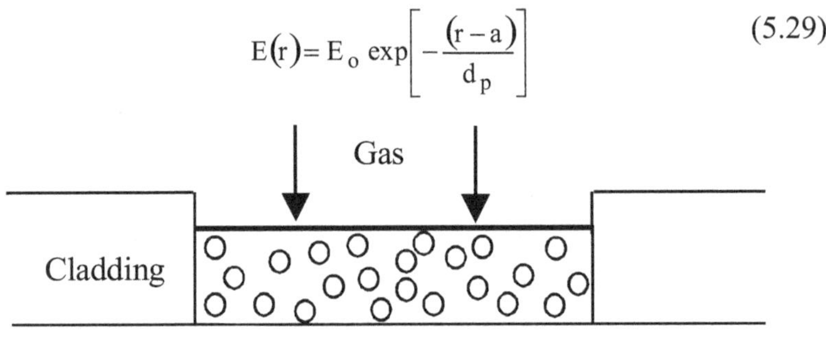

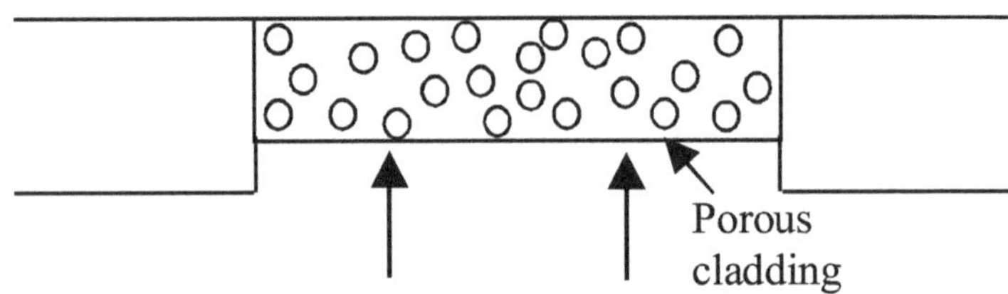

Figure 5.13: Fiber optic evanescent field absorption probe with porous cladding for the detection of a gas.

where d_p is the penetration depth as defined by Eq. (5.2). When gas comes in contact with the fiber it diffuses through the porous cladding and interacts with the evanescent field present there. Since the diffusion process is time-dependent, the evanescent absorbance will be time-dependent. As we know that an absorbing cladding is characterized by a complex refractive index, $n_2 + in^*_2$, the power lost from the field in a volume V is proportional to[14]

$$\Delta P = \int_V n_2 n_2^* |E^2| dV \tag{5.30}$$

The imaginary part of the refractive index (n^*_2) of the porous cladding is related to the bulk absorption coefficient (α) through the following relation

$$\alpha = \frac{4\pi n_2^*}{\lambda} \tag{5.31}$$

For a cylindrical volume, $dV = r dr d\phi dz$, the power loss in a length dz of the fiber is proportional to dz as well as guided core power (E_o^2). Thus the evanescent absorbance is directly proportional to

$$\int \alpha(r,t) \exp\left[-2(r-a)/d_p\right] r dr d\phi$$

According to Beer Lambert law, $\alpha(r,t)$ is directly proportional to the concentration of the absorbing species, C(r,t). Therefore, the time dependent normalized evanescent absorbance can be written as

$$\frac{A(t)}{A(\infty)} = \frac{\int_a^b \int_{\theta_1}^{\theta_2} \frac{C(r,t)}{C(\infty)} \exp\left[-2(r-a)/d_p\right] r dr d\theta}{\int_a^b \int_{\theta_1}^{\theta_2} \exp\left[-2(r-a)/d_p\right] r dr d\theta} \tag{5.32}$$

where A(t) is the evanescent absorbance at time t after the gas begins to diffuse into the porous cladding, $A(\infty)$ is its saturation value i.e. the absorbance when a uniform concentration of the gaseous absorber exists throughout the cladding, C(r,t) is the concentration of the gas in porous cladding at a distance r from the axis of the core at time t, $C(\infty)$ is its saturation value, θ is the incident angle with respect to the normal to the core-cladding interface and (θ_1, θ_2) is the range of incident angle. If the Fickian diffusion into a cylindrical surface of radius b is assumed then the concentration C(r,t) of the gas in porous cladding is given by

$$C(r,t) = C(\infty)\sqrt{\frac{b}{r}}\,\mathrm{erfc}\left[\frac{b-r}{2\sqrt{Dt}}\right] \tag{5.33}$$

provided that $Dt/b^2 < 0.02$; D is the diffusion coefficient of the gas in the porous cladding. Substitution of penetration depth and Eq.(5.33) in Eq.(5.32) gives

$$\frac{A(t)}{A(\infty)} = \frac{\int_{\theta_1}^{\theta_2} \int_{a/b}^{1} \frac{1}{\sqrt{R}}\,\mathrm{erfc}\left[\frac{1-R}{2\sqrt{Dt/b^2}}\right] \exp\left[-B\left(R-\frac{a}{b}\right)\right] R dR d\theta}{\int_{\theta_1}^{\theta_2} \int_{a/b}^{1} \exp\left[-B\left(R-\frac{a}{b}\right)\right] R dR d\theta} \tag{5.34}$$

where

$$R = r/b,$$

and

$$B = \frac{4\pi n_1 b}{\lambda}\left(\sin^2\theta - \sin^2\theta_c\right)^{1/2}$$

Integrals in Eq.(5.34) cannot be evaluated analytically. Therefore, to see the variation of normalized evanescent absorbance with dimensionless parameter (Dt/b^2) integrals have to be evaluated numerically. The variation obtained for the following values of the parameters is shown in Fig. 5.14: n_1=1.50, NA = 0.6, a = 140 μm, b=150 μm, λ = 3.36 μm, θ_1 = 66.5° and θ_2 = 90°.

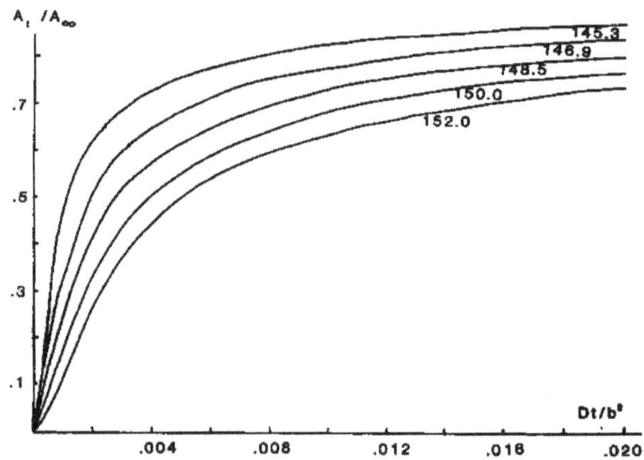

Figure 5.14: Variation of normalized evanescent absorbance with Dt/b^2. Reprinted with permission[13].

It can be seen that as the time increases the evanescent absorbance increases. This is due to the diffusion of gas molecules into the porous cladding. The rate of increase in absorbance increases with the increase in diffusion coefficient or with the decrease in the radius of the cladding. The fluoride glass core fiber with Teflon cladding which is porous was used for the detection of propane[13]. For the sensor the protective acrylate coating is removed from a 12 cm length of the fiber. This acts as the sensing length and is enclosed in a flow-through gas cell.

The porous cladding improves the evanescent field but reduces the actual volume of the gas present in the evanescent field. This reduces the improvement. Thus these points must be taken into account when assessing the actual improvement achieved. To overcome this, D-shaped fiber (see Fig. 2.9) is used for the detection of methane gas[15]. The cross-sectional shape of the D-fiber allows access to the evanescent field from the core along extended fiber lengths. The D-fiber is a single mode fiber and hence gives high sensitivity and better stability than multimode fiber. Since the sensitivity of the sensor increases with the increase in the length of the fiber, a long length of the D-fiber can be used to sense the gas. Since whole of the D-fiber is a sensing fiber it will be suitable for distributed sensing using time resolved interrogation of the gas concentration along the fiber length (see chapter 13).

The detection of gases is important for

(i) environmental and industrial process monitoring in spill (alarm, level) and continuous (control) modes

(ii) emission control in incinerator and boiler stacks, process gas monitoring

(iii) air pollution control, monitoring of gas scrubbing and combustion, exhaust gas analysis, monitoring of parking garages and tunnels

(iv) on line remote spectroscopy of chemical gas reactions in chemical plants or in autoclave or vessels of laboratories.

5.13 Hydrocarbons detection in water

There are number of cases in which the analytes and solvents have strong absorption peaks in the same wavelength region. In such cases it becomes difficult to separate the absorption of light by analytes and solvent. One such example is chlorinated hydrocarbons (CHC) in water. The CHC compounds exhibit their strongest absorption bands in the midinfrared (MIR) region (> 10 μm) and the strong absorption bands of water also lie in that region. Therefore it becomes difficult to find the amount of chlorinated hydrocarbons using absorption technique. Chlorinated hydrocarbon solvents have been widely used in chemical and industrial processes, *e.g.* in the production of PVC, as solvents and as degreasing agents for metal surfaces. Large amounts of these substances have been spilled in the environment and are now contaminating soil, drainage and ground waters. Organochlorides, mainly trichloro- and tetra-chloroethylene, are major contaminants in ground, surface and drinking water and are mostly toxic or even carcinogenic. To assess the pollution of natural surroundings, the knowledge of the concentration of these pollutants is very important. There are several analytical techniques (chromatographic methods) to quantify these contaminants. However these techniques are not very sensitive. Further these are not suitable for in situ or on site measurements. Optical fiber sensors have been developed to remove these disadvantages.

A fiber-optic sensor based on evanescent field absorption for the detection of chlorinated hydrocarbons (CHC) in water has been reported in the literarure[16,17]. As mentioned above, the CHC compounds exhibit their strongest absorption bands in the mid-infrared (MIR) region (> 10 μm). If absorption spectroscopy is to be used for their detection then PCS fibers cannot be used because these are not transparent in MIR region. Silver halide fibers are more suitable for the detection of CHC because of wide spectral range. In spite of better transparency in silver halide fibers the sensitivity of the sensor is found to be very low. This is because of the presence of strong absorption bands of water in that region. As mentioned above these absorption bands interfere with the measurements. Thus to achieve high sensitivity the CHC must be concentrated within the region adjacent to the fiber, at a depth to which the evanescent wave penetrates. Further, the absorption of evanescent field by water should be excluded from the measurements. This is done by coating the silver halide fiber with low density polyethylene (LDPE) which has good sorption properties for CHC and has only two weak absorption bands in the region of interest. The probe is prepared by dipping the fiber into a solution of 3 wt/wt % LDPE in decalin. After 1 hour the fiber is withdrawn by hand and air dried. A film of thickness more than the penetration depth of the evanescent wave is obtained by this method. The refractive index of LDPE film is about 1.52 which is less than that of the fiber core (2.06). This satisfies the wave guidance condition.

When the coated fiber is dipped into water with chlorinated hydrocarbons the CHC molecules diffuse out of the aqueous phase into the polymer while the water is kept away from the infrared beam launched into the fiber as shown in Fig. 5.15. The CHC is enriched near the fiber and the water is excluded from it. Thus the sensitivity of the sensor is increased. The infrared light is launched into the fiber and is detected at the end of the fiber by HgCdTe (MCT) detector. Using this technique, trace amounts of chlorobenzene, trichloroethylene and tetrachloroethylene can be detected.

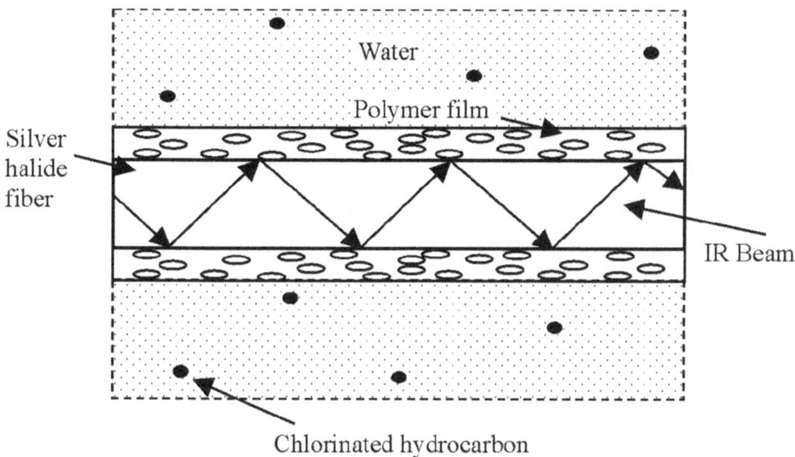

Figure 5.15: Fiber optic probe for the detection of CHC in water[16,17].

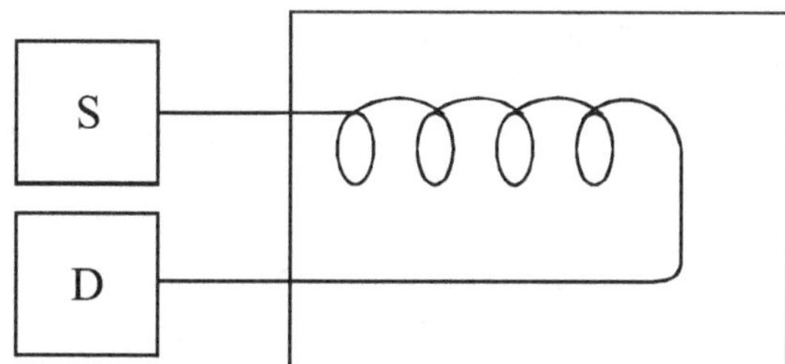

Figure 5.16: Coiled probe of a fiber optic evanescent field absorption sensor for the detection of CHC.

In another sensor a quartz glass fiber with a polysiloxane cladding is used as *in situ* measuring probe[18]. The siloxane cladding acts as a selective membrane that enriches non-polar organic compounds due to hydrophobic properties. Hence, interactions of the evanescent field at the core-cladding interface with organic species penetrating into the cladding can be measured without interferences from the broadband water OH absorption. Their probe design is different from that of Fig. 5.15. It is similar to that shown in (Fig. 5.16). Apart from acting as membrane the advantage of this kind of probe is the increase in the evanescent field which increases the sensitivity. The nylon jacket from the fiber is dissolved by boiling the coiled sensor in propylene glycol heated to 165°C for 30 min. The sensor response can be determined by measuring the output power for different amounts of $CHCl_3$ dissolved in water. A linear relationship is obtained between evanescent absorbance and the concentration in the range 80-6800 mg/l. The response time of the sensor[18] is 5-10 min.

References

1. B.D. Gupta, C.D. Singh and A. Sharma (1994) Fiber-optic evanescent field absorption sensor : Effect of launching condition and the geometry of the sensing region. *Opt. Engn*. **33**, 1864-1868.
2. S.K. Khijwania and B.D. Gupta (1999) Fiber optic evanescent field absorption sensor : Effect of fiber parameters and geometry of the probe. *Opt. Quant. Electron*. **31**, 625-636.
3. B.D. Gupta and S.K. Khijwania (1998) Experimental studies on the response of the fiber optic evanescent field absorption sensor. *Fiber Integrat. Opt*. **17**, 63-73.
4. B.D. Gupta, A. Sharma and C.D. Singh (1993) Evanescent wave absorption sensors based on uniform and tapered fibers : A comparative study of their sensitivities. *Int. J. Optoelectron* **8**, 409-418.
5. B.D. Gupta and C.D. Singh (1994) Fiber-optic evanescent field absorption sensor : A theoretical evaluation. *Fiber Integrat. Opt*. **13**, 433-443.
6. S.K. Khijwania and B.D. Gupta (1999) Fiber optic evanescent field absorption sensor based on tapered probe: Effect of fiber parameters on the response curve. *Proc. SPIE* **3666**, 578-584.
7. B.D. Gupta, A.K. Tomar and A. Sharma (1995) A novel probe for an evanescent wave fiber-optic absorption sensor. *Opt. Quant. Electron*. **27**, 747-753.
8. B.D. Gupta, H. Dodeja and A.K. Tomar (1996) Fiber optic evanescent field absorption sensor based on U-shaped probe. *Opt. Quant. Electron*. **28**, 1629-1639.
9. S.K. Khijwania and B.D. Gupta (2000) Maximum achievable sensitivity of the fiber optic evanescent field absorption sensor based on U-shaped probe. *Opt. Commun*. **175**, 135-137.

10. V. Ruddy, B.D. MacCraith and J.A. Murphy (1990) Evanescent wave absorption spectroscopy using multimode fibers. *J. Appl. Phys.* **67**, 6070-6074.

11. J. Heo, M. Rodrigues, S.J Saggese and G.H. Sigel (1991) Remote fiber-optic chemical sensing using evanescent-wave interactions in chalcogenide glass fibers. *Appl. Opt.* **30**, 3944-3951.

12. K. Taga, B. Mizaikoff and R. Kellner (1994) Fiber optic evanescent field sensors for gaseous species using MIR transparent fibers. *Fresenius J. Anal. Chem.* **348**, 556-559.

13. V. Ruddy and S. McCabe (1990) Detection of propane by IR-ATR in a Teflon-clad fluoride glass optical fiber. *Appl. Spectrosc.* **44**, 1461-1463.

14. A.W. Snyder and J.D. Love (1983) Optical waveguide theory. Chapman & Hall, London, p. **598**.

15. B. Culshaw, F. Muhammad, G. Stewart, S. Murray, D. Pinchbeck, J. Norris, S. Cassily, M. Wilkinson, D. Williams, I. Crisp, R. Van Ewyk and A. McGhee (1992) Evanescent wave methane detection using optical fibers. *Electron. Lett.* **28**, 2232-2234.

16. R. Krska, E. Rosenberg, K. Taga, R. Kellner, A. Messica and A. Katzir (1992) Polymer coated silver halide infrared fibers as sensing devices for chlorinated hydrocarbons in water. *Appl. Phys. Lett.* **61**, 1778-1780.

17. R. Krska, R. Kellner, U. Schiessl, M. Tacke and A. Katzir (1993) Fiber optic sensor for chlorinated hydrocarbons in water based on infrared fibers and tunable diode lasers. *Appl. Phys. Lett.* **63**, 1868-1870.

18. J. Burck, J.P. Conzen and H.J. Ache (1992) A fiber optic evanescent field absorption sensor for monitoring organic contaminants in water. *Fresenius J. Anal. Chem.* **342**, 394-400.

6

Reagent-Mediated Sensors

6.1 Introduction

There are a variety of intrinsic and extrinsic fiber-optic sensors which use absorption spectroscopy indirectly for the detection of various chemical species. These include pH, metal ions, oxygen in water *etc.* In such sensors a suitable reagent is immobilized on a solid support. The support can be the core of the fiber or some optical material attached at the end of the fiber. Further it is optically transparent to allow transmission of the light signal from the interface and should remain inert to the chemical reaction being analyzed. The analyte (or the chemical species) when comes in contact of the reagent affects the absorption properties of the reagent. In other words, in these sensors, the reagent acts as a chemical transducer for analytes that are not directly measurable by optical techniques. Following conditions for the use of reagents in sensors must be met:

1. The reagent must be highly sensitive; a large change in the absorption spectrum of the reagent should occur for small change in the concentration of the analyte.
2. The change in absorption spectrum should be easily detected.
3. The reagent concentration must be very small to avoid changing the concentration of acceptor and donor. This implies that a high molar absorption coefficient is desirable for a reagent.
4. The reagent should be specific for a given analyte.
5. Sometime the choice of the reagent is restricted by the availability of the light source and the detector. The absorption peak of the reagent chosen should be close to the wavelength of the light source. Also the detector should be sensitive at that wavelength.

Most of the available LED's and PIN photodiodes cover a limited wavelength range (>550 nm). For low wavelength use blue LED along with a PMT or avalanche photodiode can be used. But these are most costly and have problems of long-term drift and temperature stability. Thus reagents with a maximum absorbance at wavelengths less than 500 nm are not suitable for industrial applications.

In this chapter we shall describe the sensor configurations, immobilization techniques and some of the reagent-mediated fiber optic sensors utilizing absorption spectroscopy.

6.2 Sensor configurations

There are two kinds of reagent mediated fiber optic sensors. The one in which the reagent is immobilized on the surface of the fiber core or at the end face of the fiber. In the second, the reagent is immobilized on some solid support (such as some small solid spheres) separately and then placed at one of the ends of the fiber. Typical reagent-mediated sensor configurations are shown in Fig. 6.1. In Fig. 6.1 (a) the reagent

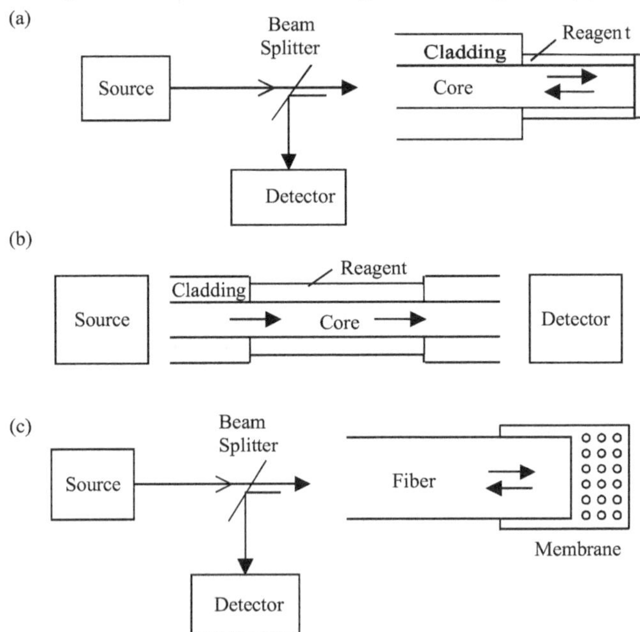

Figure 6.1: Different sensor configurations.

is immobilized onto the core of the fiber at one end and the end is made reflecting. The refractive index of the film with entrapped reagent is smaller than that of the fiber core. When light is coupled to the fiber it reaches the sensing region. Here the evanescent field interacts with the reagent in the presence of analytes. The light after interaction is reflected back from the reflecting fiber end and again interacts with the reagent. The returned light is detected by means of a power meter after separating it from the incident light by using a beam splitter. In Fig. 6.1 (b) the reagent is immobilized onto the core in the middle of the fiber. The evanescent field after interaction with reagent in the presence of analytes is detected by a power meter kept at the other end of the fiber. In Fig. 6.1 (c) the reagent is immobilized on the micron size spheres and then placed at the end of the fiber using an analyte permeable membrane. The light exited from the fiber end interacts with these spheres. The absorption and scattering of light occurs in the presence of analytes. The scattered light coupled to the fiber is detected at the other end of the fiber similar to configuration (a).

6.3 Immobilization methods

Surface modification such as mechanical polishing, electrochemical modification and chemical etching of the surface of the substrate is required before immobilization. The immobilization of most reagents results in a small change of their spectral characteristics. It is therefore required to determine the respective data of the immobilized reagent rather than to use the values determined in bulk solution. The changes in spectral characteristics occur due to interactions between reagents and polymer surface. There are a large number of immobilization methods used for the fabrication of reagent-mediated fiber optic sensors. These are physical entrapment, adsorption, electrostatic, covalent and sol-gel methods. We, briefly, describe these below.

6.3.1 Physical entrapment

In the physical entrapment, reagent is mixed directly in the polymeric support in a silica gel or in a solution that is subsequently polymerized. The molecules are incorporated in the interior of the polymer. A common technique is to trap a reagent in microspheres with diameters less than 1 μm. The technique is simple and effective but reproducibility from mixing is generally poor.

6.3.2 Adsorption

Adsorption is a physicochemical immobilization due to weak electrostatic and van der Waals forces of interaction between the reagent and the support. Reagent immobilization can be done by polymerization or adsorption on a preactivated surface. It is a simple technique. In the method, reagent is dissolved in an alcohol. The end of the fiber with polar cross-linked polymer or copolymer is immersed into the reagent solution. It is then washed to remove unadsorbed reagent.

6.3.3 Electrostatic (ionic) binding

A polymer containing charged groups can be bound to oppositely charged ions by electrostatic forces. The charged polymers are immersed in an oppositely charged alcohol solution having reagent. Due to electrostatic interaction the binding of the reagent to the polymer surface occurs. The loading conditions for reagent, such as the immersion time, must be closely controlled to obtain reproducible sensors. The major advantage of electrostatic immobilization is the simplicity of the procedure. The film thickness, in this case, depends on the immersion time.

6.3.4 Covalent immobilization

In this method, the immobilization is accomplished by creating a covalent bond between the reagent and a polymer surface. For this, one or two activation steps are usually required to make the reagents undergo a facile room temperature reaction. The first step usually involves the modification of the polymer to provide it with sufficiently reactive function. A similar procedure may be required when the reagent does not possess chemical functions suitable for immobilization.

6.3.5 Sol-gel

Sol-gel method is a novel process for the preparation of ceramics and glasses. It has several advantages over other methods of film deposition. Apart from simplicity, the film produced by sol-gel process is tough, inert, intrinsically bound to the fiber core and more resistant than polymer films in aggressive environments. The sol-gel method can be divided into two steps. The first step is the hydrolysis of metal alkoxide to produce hydroxide. The second step is the condensation polymerisation of hydroxyl group. The process is generally conducted at low temperatures and therefore it is possible to immobilize organic macromolecules in inorganic glass networks. The advantages of sol-gel glass as a solid support for chemical reagents in optical sensor applications are the following:

(i) The low temperature processing allows the use of low thermal stability dyes for the sensing.

(ii) The glasses are chemically inert, photostable and thermally stable compared with a plastic matrix. Thus they are highly suitable for applications in harsh environments or in the food industry.

(iii) In the glass film reagents are non-leachable thus it has clear advantage over reagent adsorption technique.

(iv) The sol-gel technique can be applied to any organic indicator having a molecular size larger than that of the glass former which forms the cage structure.

(v) The sol-gel processing technique allows the sensor designer to tailor the properties of the coating layer, such as pore size distribution, film thickness and refractive index, to the requirements of a particular sensor.

The metal alkoxide *i.e.* the precursor most widely used is tetraethyl orthosilicate (TEOS), $Si(OC_2H_5)_4$. To perform the hydrolysis reaction of TEOS, it is thoroughly mixed with deionized water and ethyl alcohol. Sometime hydrochloric acid is also mixed to increase the rate of hydrolysis. The solution is kept for few hours for the condensation process. The condensation liberates small molecules, such as water or alcohol. This type of reaction can continue to build up larger and larger molecules by the process of polymerisation. Dilute solution containing these polymerised species is then used to coat the substrates. The reagent, if used, is incorporated at the precursor solution stage.

The thin films are produced on the substrates by using spinning or dip coating method. In dip coating method, substrate is lowered into the still ungelled coating solution and then withdrawn at a definite speed. The polymeric species are concentrated on the substrate surface by gravitational draining accompanied by vigorous evaporation and further condensation reactions. The condensation and polymerisation during deposition stage lead to the formation of microscopic clusters in three dimensions, increased solution viscosity and eventually, formation of a gel. The gel is an amorphous, porous material with the liquid solvents (ethyl alcohol and water) still contained within the pores. Low temperature curing expels all the liquids and leaves the porous oxide. Further curing at higher temperatures increases the density, strength, hardness and the refractive index of the film. There are several factors which affect the coating process. These are composition and pH of the precursor liquid, coating speed, temperature, pressure and humidity. In the following sections we shall describe a number of reagent-mediated fiber optic sensors.

6.4 Humidity sensor

Humidity is generally reported as relative humidity (RH). It is defined as the ratio of concentration of water molecules in the given volume of the environment to the saturated concentration of water molecules in the same volume at a particular temperature and pressure. The device that measures the relative humidity is called hygrometer. The conventional hygrometers are typically based on electrical properties of the sensing materials. These can be divided into two categories, the resistive and capacitive types. For the former, the surface electrical conductivity or dielectric constant of porous metal oxides is enhanced by adsorption of the surrounding water vapour molecules owing to a change in either electronic or ionic conduction. For the latter, the capacitance is enhanced by the absorption of water and thus causes a variation in the impedance of the sensing material. Materials that have this property include polyimide, polyvinyl alcohol and cellulose. In recent years there has been great interest in the development of fiber optic humidity sensors which are independent of electrical and magnetic interference and potentially capable of being used at high temperatures. Although number of techniques and reagents have been used to fabricate fiber optic humidity sensors we shall here describe some of these.

A fiber optic probe utilizing cobalt chloride as reagent to sense humidity is shown in Fig. 6.2. The cobalt chloride changes colour when exposed to gaseous moisture at room temperature. Thus if the light of an appropriate wavelength is passed through the cobalt chloride its absorption will depend on the amount of moisture present around cobalt chloride. To fabricate the probe, an optical fiber with 0.5 cm long porous core section was used[1]. It was prepared by chemically leaching a small section (1 cm) of a phase-separated sodium borosilicate glass fiber. The porous part of the fiber was immersed in a cobalt chloride aqueous solution. The fiber was then dried in dry air overnight at room temperature. After drying it was placed in a small flowing gas chamber. A digital hygrometer with 0.1% resolution was used to monitor the humidity in the gas chamber. Light from a quartz halogen lamp was focused onto one of the ends of the fiber after passing through a monochromator. The transmitted light was picked up by a photodiode. Most of the light launched into the porous fiber is influenced by the indicator absorption, some is scattered out by the porous structure of the glass host. Since the core is absorbing, it just acts as the light is passing through the sample absorption cell. Thus it can be called in-line optical absorption. In the presence of saturated amount of moisture (highest humidity), $CoCl_2$ has a absorption peak around 500 nm and appears pink. When it is dried well it changes colour to bright blue and has high optical absorption between 550 and 750 nm. The maximum change in absorption occurs near 690 nm. Therefore 690 nm wavelength was used as the monitoring wavelength for recording the humidity-induced intensity changes. When the moisture is admitted into the chamber, the transmitted intensity increases rapidly and after a couple of minutes the output signal approaches a constant value. The response time of this sensor is about 2-3 min while the dynamic range of the sensor depends on the concentration of the cobalt chloride present. By increasing the concentration the measurement range can be increased. The sensor is reversible and is reported to be capable of detecting relative humidity down to 0.5% at 25°C.

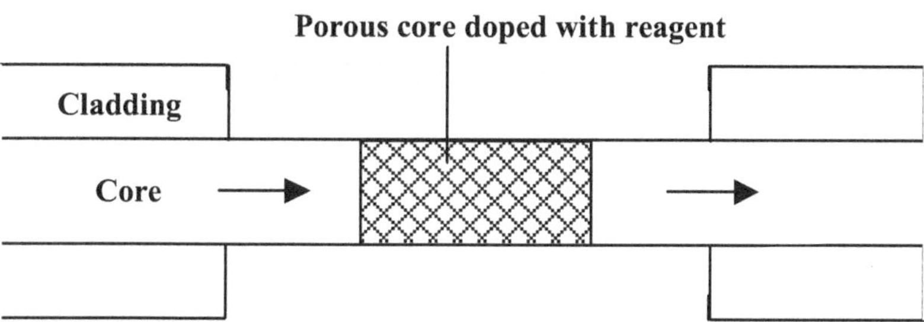

Figure 6.2: Porous glass optical fiber humidity probe.

In many industries engaged in activities such as food processing and storage, gas and power transmission, materials synthesis and device fabrication the humidity measurement at high temperatures is required. The humidity sensor described above cannot be used for temperatures above 100°C. Therefore one needs a humidity sensor which can work at temperatures greater than 100°C so that it can be used for the applications mentioned above. A humidity sensor which uses porous glass fiber in conjunction with an indicator, cobalt sulphate ($CoSO_4$), was reported[2] to measure moisture level at temperatures above 100°C. Except indicator the sensor is similar to that described above[1]. The moisture concentration is monitored at 670 nm wavelength. The response time of the sensor ranges from 1 to 3 min.

A humidity sensor based on evanescent field absorption was also reported around the time when the above mentioned high temperature humidity sensor was reported. If a polymethylmethacrylate (PMMA) film containing phenol red (PR) dye is exposed to moisture environments, its absorption coefficient around 530 nm changes remarkably. This fact was used for the fiber optic humidity sensor utilizing evanescent field in conjunction with PR doped PMMA film[3]. A PR doped PMMA film of about 1 mm thickness was prepared on the core of the plastic fiber of 1 mm core diameter by means of a spinner. The schematic of the sensor is shown in Fig.6.3. A moisture environment around the sensing fiber was produced in a thermally controlled water pot and is injected into the flow cell. Humidity in the flow cell was measured by means of an electronic hygrometer. A green He-Ne laser operating at 544 nm wavelength was used to launch light into the fiber. The transmitted light was measured by means of a photomultiplier tube (PM) for various values of atmospheric humidity. The calibration curve of the sensor is shown in Fig.6.4. As can be seen from the figure, the transmitted light intensity decreases exponentially with increasing humidity. The temperature effect on the humidity sensing has been reported to be small while the response of the sensor is fast.

Based on the absorption of light by PR doped PMMA film in the presence of moisture, a highly sensitive fiber optic humidity sensor was developed in our laboratory[4]. A PMMA film containing phenol red dye is deposited over the unclad core of a highly multimode plastic clad silica (PCS) fiber. The refractive index of the PMMA film was greater than that of the fiber core. When the light launched in the fiber reaches the coated portion of the fiber a fraction of light guided in the core transmits to the film after each reflection at the core-film interface as it propagates along the fiber axis. If the wavelength of the light launched in the fiber is close to the peak absorption wavelength of the phenol red dye present in the film the absorption of the transmitted light in the film occurs. The absorption depends on the moisture present in the air around the phenol red doped PMMA film. After traversing the film, the light couples back to the fiber core and is detected by a power meter at the other end of the fiber. For a given length of the

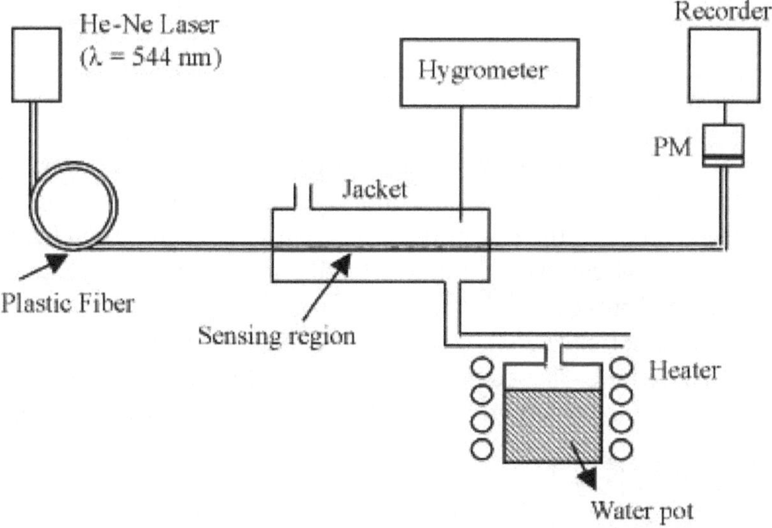

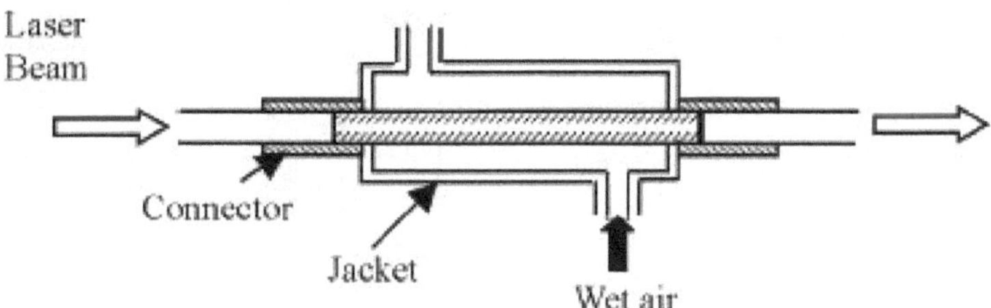

Figure 6.3: Schematic diagram of a humidity sensor utilizing dye doped plastic fiber. Redrawn[3].

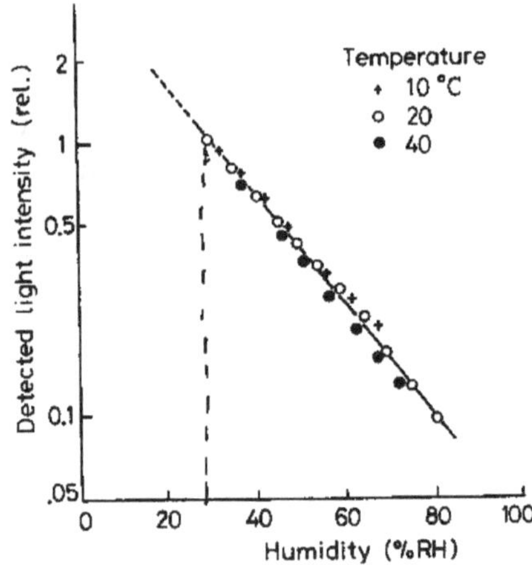

Figure 6.4: Transmitted light intensity as function of atmospheric pressure Reprinted with permission[3].

coated probe, the amount of absorption depends on the angle of incidence of the ray at the core-film interface. Smaller the angle of incidence larger is the transmission of light in the film. The angle of incidence of the ray at the core-PMMA film interface can be decreased if a U-shaped probe is used. The U-shaped probe is fabricated according to method described in Sec. 5.8. The experimental set up for a humidity sensor based on U-shaped probe[4] is shown in (Fig. 6.5). The unclad U-shaped probe is coated with phenol red doped PMMA film and then

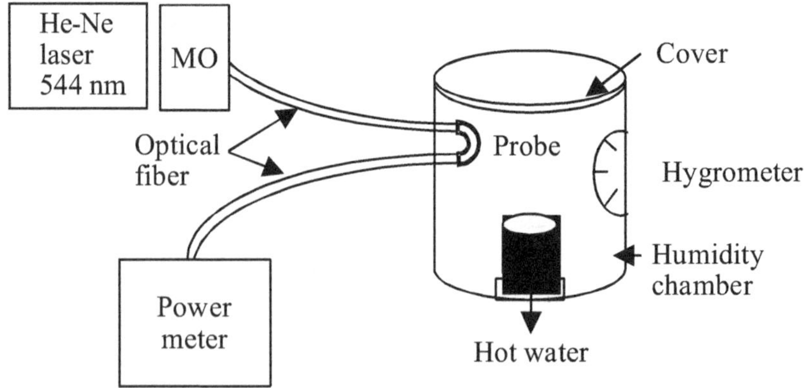

Figure 6.5: Fiber optic humidity sensor based on U-shaped probe. Reprinted from ref. 4 with permission from Elsevier.

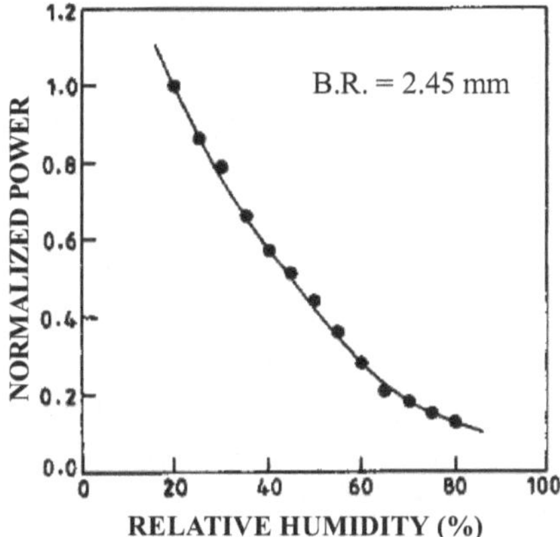

Figure 6.6: Calibration curve of the fiber optic humidity sensor for the U-shaped probe of 2.45 mm bending radius. Reprinted from ref. 4 with permission from Elsevier.

fixed inside a plastic cylindrical container (humidity chamber). A hygrometer is also placed inside the chamber to calibrate the sensor. Light from a He-Ne laser operating at 544 nm is focused using a microscope objective on the input face of the fiber. The output power is measured as a function of humidity inside the chamber. The calibration curve of the sensor is shown in Fig. 6.6. Its operating range is from 20 to 80% RH and its response time is observed to be about 5 sec. The sensitivity of the sensor increases as the bending radius of the probe decreases as shown in Fig. 6.7.

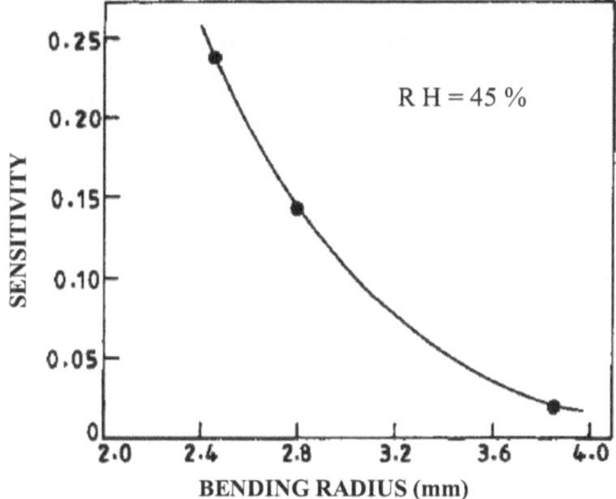

Figure 6.7: Variation of sensitivity with bending radius of the U-shaped probe for 45% RH. Reprinted from ref. 4 with permission from Elsevier.

6.5 pH sensor

The measurements of pH are very important in many chemical and industrial processes, in environmental control and the biomedical field. The conventional method used to measure pH is electrometric. It uses glass electrodes and is based on the measurement of electromotive force of a cell which has a reversible electrode whose electrochemical potential depends on the activity of the hydrogen ions and hence on the pH. The use of such an electrode for pH measurement is limited. In medicine where on-line monitoring of blood pH is desirable, the use of glass electrode proved not to be practical because of their large size, rigid design and possible electrical hazard. In addition, the presence of local electrical potentials in the body may affect the performance of the pH electrodes. Further, pH electrodes cannot be used for solutions through which an electrical current is flowing, for example in electrolysis cells and batteries. Other problems include susceptibility of wire leads and couplings to deterioration under corrosive conditions, or conditions of altering temperatures.

Many of the problems can be overcome if optical fibers along with spectroscopic techniques are used for sensing. Optical fiber sensors offer several advantages over glass electrodes. The important ones are immunity to electromagnetic interference, their minute size and flexibility. The small size and flexibility are important for invasive analysis of plant and animal tissues.

In old days pH was defined as the negative logarithm of the hydrogen ion concentration $[H^+]$. Today, the actual acidity of a solution is related to the activity of the hydrogen ions a_{H+}. Thus, the new definition of pH is

$$pH = - \log a_{H+} \tag{6.1}$$

Activity and concentration are related by the activity coefficient f_H+ through the relation

$$a_{H+} = f_{H+} [H^+] \tag{6.2}$$

In optical sensors light is passed through an optical fiber to the active end of the sensor where it interacts with the chemical indicator in the presence of fluid sample, which alters the intensity of the beam by absorption or fluorescence. The modified optical signal is guided by the same or another fiber to the detector. The pH indicator is usually confined at the fiber end or immobilized on the surface of the fiber core. The dissociation constant of the pH indicator when it comes in contact of water is defined by the following equations

$$HA + H_2O = H^+ + A^- \tag{6.3}$$

$$K = \frac{a_{H^+} C_{A^-}}{a_{H_2O} C_{HA}} \frac{f_{A^-}}{f_{HA}} \tag{6.4}$$

where C_{A-} and C_{HA} are the concentrations of the indicator conjugate base and conjugate acid, respectively while f_{A^-} and f_{HA} are the corresponding activity coefficients. In logarithmic form it can be written as

$$pH = pK + \log\left(C_{A^-} / C_{HA}\right) + \log\left(f_{A^-} / f_{HA}\right) - \log a_{H_2O} \tag{6.5}$$

The above equation is called Henderson-Hasselbalch equation. In Eq. (6.5), the last term can be ignored in a solution in which the ionic strength is < 4 M. The activity coefficient is related to the ionic strength (I), according to the specific interaction theory, by the relationship

$$\log f_i = -z_i^2 D + \sum_k \varepsilon(j, k, I) m_k \tag{6.6}$$

where

$$D = 0.51\left(I^{1/2}\right) / \left[1 + a\left(I^{1/2}\right)\right] \tag{6.7}$$

z_i is the charge on the ion i, and ε is the interaction term between the ion i and the ions, with the sum extended to the k ions present in the medium. Parameter 'a' in Eq. (6.7) is related to the ionic size and for most ions is taken to be 1.5. The value of the specific interaction term, second term in Eq. (6.6), depends on the type of the indicator used and on the kinds of ions j, k present in the medium. It is thus apparent from Eq. (6.5) that pH is strictly dependent on the ionic strength of the sample to be investigated. The third term of Eq. (6.5) can be ignored only in the case of a very dilute solution because the activity coefficient can be assumed to be equal to unity. But this is not the case in real situations.

Apart from ionic strength pH also depends on the temperature of the sample. The variation of pK with thermodynamic temperature T, in Eq. (6.5), can be expressed by the equation of the form

$$pK = \frac{A}{T} + B + C.\log T \tag{6.8}$$

where A, B and C are constants which may be obtained from the standard changes of enthalpy (ΔH^o), entropy (ΔS^o) and heat capacity (ΔC_p^o) for the dissociation process.

In optical pH measurements, the effect of ionic strength on the dissociation constant results in a shift of the pH-dependent dissociation curves. Therefore, ionic strength effects are indistinguishable from signal changes caused by pH. The pH error caused by ionic strength depends on the type of indicator used and on the concentrations of all ionic species in solution. The resulting errors are particularly significant when the composition of the sample solution differs significantly from that of the calibration solution or when it changes during measurements. Therefore, optical pH sensor cannot be applied to unknown sample solutions to give a reliable pH indication. Further, the calibration curve of the pH sensor will change if the temperature of the solution changes resulting in giving wrong pH value.

Optical fiber pH sensors are based on pH-induced reversible changes in optical or spectroscopic properties such as absorbance, reflectance, fluorescence, energy transfer *etc*. The major component in an optical fiber pH sensor is a pH sensitive layer/film that is prepared by immobilizing a pH sensitive dye onto the tip or sides of an optical fiber. To immobilize, many research groups have used either covalent chemical linking or simple physical encapsulation techniques. For covalent linking of the dye molecules surface modification is required. The covalent linking method described in Sec. 6.3 is a long, complex and tedious method and may lead to loss of dye sensitivity or result in poor absorption and fluorescence properties. In the non-covalent immobilization techniques, the dye is immobilized on the polymer support and entrapped behind semi-permeable membranes. The disadvantage of the non-covalent immobilization methods is that these suffer from leachability of the dye that makes the long term use of the sensor impractical. These shortcomings can be overcome if sol-gel method is used for the immobilization of pH-sensitive dyes. The pH probes are calibrated either using buffer solution of different pH or deionized water of different pH prepared by using HCl and NaOH. In this section we shall describe pH sensors based on absorption phenomenon. The fluorescence-based pH sensors will be described in chapter 7.

In 1980, the first fiber-optic pH sensor was reported[5]. Its principle was based on the pH-dependence of light absorption by phenol red, a reversible indicator. In the case of phenol red, the absorption of light increases with the increase in the pH of the dye solution in the wavelength range 500 to 600 nm. At 560 nm wavelength the change in absorption is maximum. In the probe fabricated, phenol red is encased in an appropriately permeable envelope attached at the ends of the two fibers (see Fig.6.8). One of the fibers is used to transmit light to the probe while the other fiber collects the light scattered by the dye after absorption. The intensity of the scattered light collected and guided by the second fiber is measured by means of a detector. In the probe, the dye is covalently bound to polyacrylamide microspheres. This is done to give support to the dye to prevent it from diffusing out of the envelope. The polyacrylamide microspheres also contain smaller polystyrene microspheres of approximately 1 μm diameter. These spheres provide a very effective scattering of light. All these materials are packed in an envelope of cellulosic tubing permeable to hydrogen ions. Lights of wavelengths 560 nm and 600 nm are passed through the fiber. The 600 nm light is used as the reference beam. Both the beams after multiple scattering and absorption in the packing return to the second fiber. The ratio of the intensities of two lights detected reveals the pH of the solution. The sensor is reported[5] to be capable of measuring pH in the physiological range (7.0-7.4) with an accuracy of 0.01pH. The probe diameter of the sensor is about 0.4 mm and hence can be inserted into the body using a catheter.

A fiber-optic pH sensor based on pH dependent reflectance[6] is shown in Fig.6.9. The probe consists of a bundle of 16 plastic fibers.

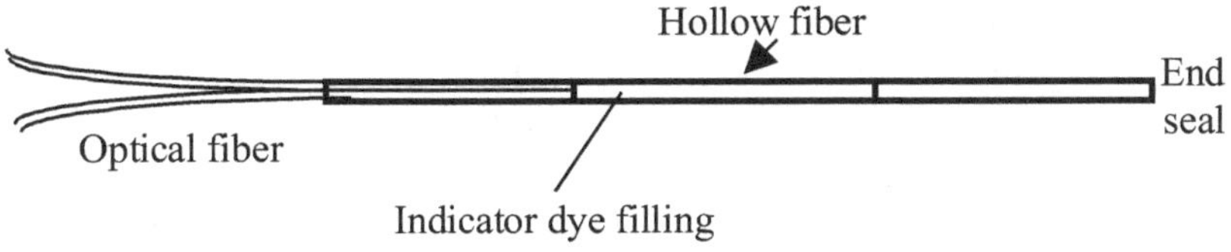

Figure 6.8: Fiber optic pH sensor. Redrawn with permission[5]. Copyright (1980) American Chemical Society.

At the tip of the bundle a styrene-divinylbenzene copolymer (XAD-2) supporting an adsorbed colorimetric pH indicator (bromothymol blue) is attached. The copolymer is retained in position using a porous membrane of polytetrafluoroethylene (PTFE). Half of the fibers of the bundle are used to transmit visible light of 593 nm wavelength to the sensitive tip and the remaining fibers are used to collect a portion of the light reflected from the polymer which is then sent to the detector. Variation in pH in the vicinity of the tip causes a variation in the attenuation of reflected light. The main disadvantage of the sensor is the significant dependence of its response on the ionic strength of the solution.

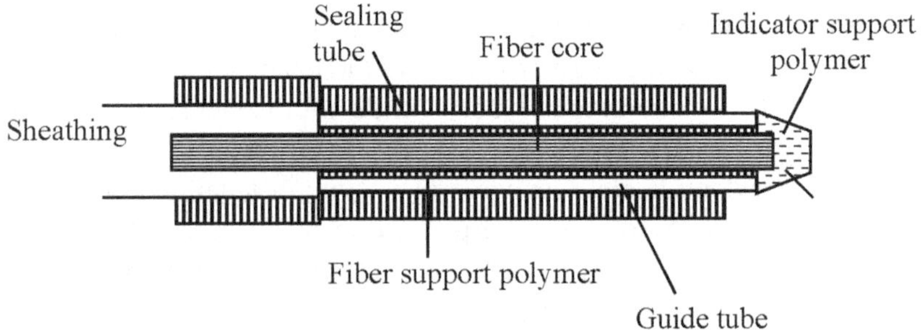

Figure 6.9: pH sensor using fiber bundle and dye adsorbed copolymer at the tip of the bundle[6]. Redrawn with permission of The Royal Society of Chemistry.

Another fiber optic pH sensor based on absorption phenomenon and fabricated by immobilizing the dye congo red to a porous cellulose triacetate film is shown in Fig. 6.10. The probe is similar to that used by Liu *et al*[7]. It consists of a fiber bundle. The central fiber is used to transmit light from a quartz-halogen filament lamp operating at 630 nm to the porous polymeric film of 0.2 mm thickness. The transmitted light is reflected back by a mirror attached at the end of the probe and again passes through the film. The reflected light is then collected by the output fibers surrounding the central fiber. The output power is detected by a photodetector. When the probe is immersed in a test solution hydrogen ions penetrate through the porous sleeves and then interact with the dye immobilized on the film. The absorption of light passing through the film depends on the pH of the test solution. As the pH increases the absorption decreases. To calibrate, the pH of the solution is adjusted by varying the amounts of HCl and NaOH in the solution. All solutions are prepared with deionized water. The ionic strength of the solution is controlled by adding NaCl salt. The dynamic range of this sensor is from 1.0 to 4.5 pH. The sensor is reported to have a rapid response time and high stability. It is immune to metal ion interferences.

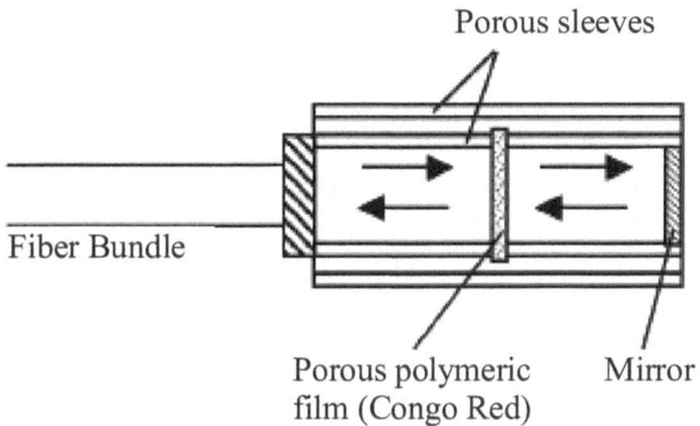

Figure 6.10: Schematic diagram of the probe for porous polymeric sensor[7].

The fiber optic pH sensors described above are extrinsic type. Now we describe few intrinsic type fiber optic pH sensors. The first intrinsic type fiber optic pH sensor was based on a porous core silica glass fiber[8]. To fabricate a pH sensitive probe bromocresol purple and bromocresol green in a 1:1 molar ratio are immobilized on the porous core using sol-gel method. Typically a 200 to 300 μm core porous glass fiber is used for the development of the sensor. The depth of the porous layer of the fiber varies from several μm to more than 100 μm. The advantage of porous fiber is that the dye gets impregnated into the pores. This increases the surface area of interaction of dye with hydrogen ions penetrating into pores. Because of high surface area of porous fiber, the light interaction is increased which leads typically to a hundred fold increase in the sensitivity of the sensor. The schematic diagram of the experimental set up and the basic configuration of the porous fiber sensing element are shown in Fig.6.11. Two hard polymer protected silica optical fiber are used to propagate light. A short piece of dye coated porous fiber is glued between these two fibers. Light from a xenon lamp and modulated by a mechanical chopper is launched into one of the ends of the fiber. A monochromator is introduced between the chopper and the fiber. The porous fiber is immersed in a cell containing buffer solution. The second fiber is connected to a photodetector. The propagating light interacts with the dye whose absorption increases with the increase in the pH of the solution. A lock-in amplifier is used to amplify the signal from the photodetector. The results are recorded by a recorder. An electronic pH meter is placed in the cell and used as a reference for the pH measurements. The sensor is stable, reversible and has short response time. Its dynamic range is from 3-9 pH unit. In spite of being highly sensitive, there has been relatively little work done on the core based sensors. This is primarily due to the difficulty of designing a structure that will allow the analyte to get into the core while maintaining the strength and transparency required of a practical optical fiber.

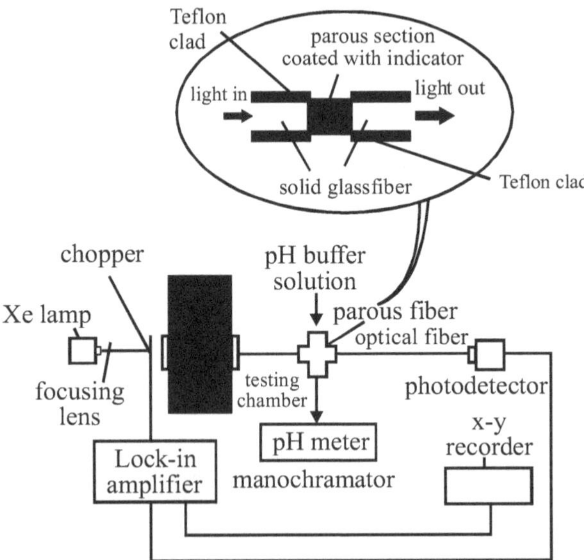

Figure 6.11: Schematic diagram of experimental set up for porous glass fiber optic pH sensor[8]. À© [1991] IEEE.

To overcome the design problem of using porous core, the porous cladding has been used for pH sensing. The porous cladding with pH sensitive dye entrapped in it is prepared over the surface of unclad core using sol-gel method[9]. The refractive index of the cladding so prepared is chosen less than that of the core. The evanescent field present in the cladding interacts with the dye in the presence of the H^+ ions. To fabricate the pH probe, the jacket and the cladding of the PCS fiber are removed from the middle portion (about few cm) of the fiber. For coating a thin film on the core of the fiber dip coating method is used. In dip coating method the unclad fiber is first lowered into the still coating solution and then it is pulled at a definite speed. The thickness of the film depends on the pulling speed. The precursor liquid used for preparing pure silica thin film on the core is tetraethyl orthosilicate (TEOS). This is used because the refractive index of the porous glass film produced is less than the refractive index of the fiber core thus allowing the wave guidance condition to be satisfied. Different dyes namely phenol red, cresol red, bromophenol blue, chlorophenol red etc are used to fabricate pH probes[9,10]. The experimental arrangement used to characterize these pH probes is shown in Fig. 6.12. Light from

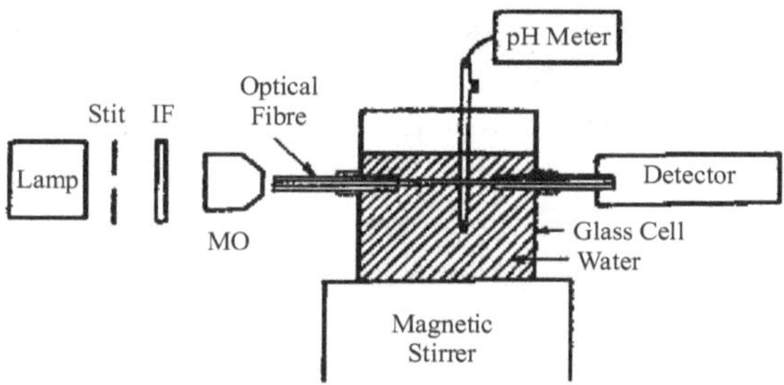

Figure 6.12: Schematic diagram of an evanescent field absorption based optical fiber pH sensor prepared by dye doped sol-gel immobilization technique. Reprinted from ref. 9 and 10 with permission from Elsevier.

a tungsten halogen lamp is launched into the coated fiber mounted in a cylindrical glass cell with the help of a circular slit and a microscope objective of numerical aperture higher than that of the fiber. The coated portion is kept in the middle of the cell. The other end of the fiber is connected to a power meter. The cell is filled with deionized water and is kept on a magnetic stirrer. To calibrate, a pH electrode is introduced in the flow cell. The pH of the water is increased or decreased by adding NaOH or HCl respectively. Interference filters of different wavelengths are introduced between the lamp and the microscope objective to scan the maximum fractional change in transmitted power over the entire pH range. The maximum fractional change in power for phenol red, cresol red, chlorophenol red and bromophenol blue occurs between 610 nm to 630 nm. Therefore, an interference filter of 620 nm wavelength is used for all the probes in the experimental arrangement. The variation of the normalized transmitted power as a function of pH of the fluid around the probe is recorded for each probe. It is observed that as the pH of the water around the coated core increases the power transmitted decreases. The modulation in power transmitted for phenol red occurs in the pH range 7.5 to 11.5. These observations are different from those obtained for phenol red in water. In water the phenol red has peak absorption around 570 nm and its absorbance changes in the pH range 6.5 to 8.0. It means that in porous glass matrix, a shift in peak absorption wavelength and an increase in the pH dynamic range occur. The dynamic ranges reported for cresol red, bromophenol blue and chlorophenol red dyes are 6.5 to 11.0, 4.0 to 7.5 and 8.5 to 13.0 respectively. The dynamic range of these and other sensors reported in the literature is about 3-4 pH unit. This is small as compared to the pH range (0-14) of the conventional pH meter. The short dynamic range of the sensor limits its applications. The dynamic range of the pH sensor can be increased by using a mixture of suitably selected pH sensitive dyes. If the mixture of cresol red, bromophenol blue and chlorophenol red are used[10] then the pH sensor prepared by sol-gel immobilization method operates in the pH range from 4.5 to 13.0 which is more than 8 pH unit. The calibration curve of the sensor developed using a mixture of dyes is shown in Fig. 6.13. The response time of the sensor is approximately 5 sec.

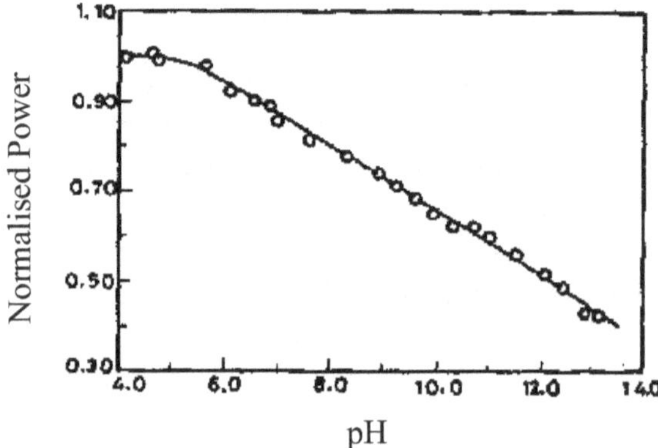

Figure 6.13: Calibration curve of the long range fiber optic pH sensor. Reprinted from ref. 10 with permission from Elsevier.

The pH sensors described above have straight probe and therefore it is not possible to use them as a point sensor. If the U-shaped probe is used for the pH sensing then it can be used as a point sensor. The experimental arrangement of the U-shaped fiber optic pH sensor[11] is shown in Fig. 6.14. A mixture of three dyes used above is selected for this sensor. Further, the method of immobilization of the mixture of dyes over the unclad U-shaped fiber is also the same as in the case of straight probe. Light from a tungsten halogen lamp is launched at the input end of the coated U-shaped fiber optic probe mounted in a cylindrical glass cell with the help of a circular slit and a microscope objective (MO). An interference filter (IF) of 600 nm is introduced between the slit and the microscope objective to achieve maximum fractional change in transmitted power per unit change in pH. The other end of the fiber probe is connected to a power-meter. The glass cell is filled with de-ionized water and is kept on a magnetic stirrer. The method of calibration is same as that of straight and uniform pH probe (Fig. 6.12). Figure 6.15 shows the variation of normalized power recorded as a function of pH of the water around the coated U-shaped probe of bending radii 3.48, 1.14 and 0.82 mm fabricated from the PCS fiber of numerical aperture 0.40. The probe with larger bending radius has small variation. In the case of probe with 0.82 mm bending radius the variation in power is large and it covers almost full range of the normalized output power in the pH range 4 to 13. If the response of this probe is compared with that of the straight probe of 5 cm length (Fig. 6.13) the decrement in power for a given pH range is more in the case of U-shaped probe. Thus the U-shaped probe with 0.82 mm bending radius is highly sensitive. The increase in bending radius decreases the sensitivity of the sensor[11].

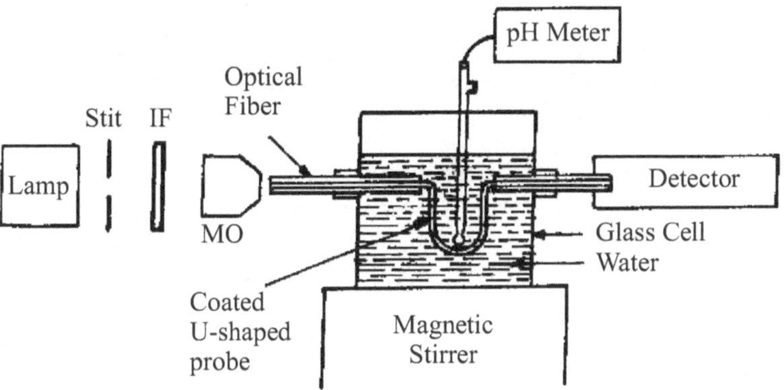

Figure 6.14: Experimental set up of the fiber optic pH sensor based on U-shaped probe. Reprinted from ref. 11 with permission from Elsevier.

Most of the pH sensors described show ionic strength dependence. The pH error caused by ionic strength depends on the type of the indicator used and on the concentrations of all ionic species in solution. The resulting errors are particularly significant when the composition of the sample solution differs significantly from that of calibration solution or when it changes during measurements. To decrease the error, an appropriate buffer for each type of sample has to be used for calibrating a pH probe. Therefore, pH sensor cannot be used for unknown sample solutions to give a reliable pH indication. The best use of pH sensors is for seawater, biofermenters, agrifood industries and other situations where the sample composition is almost invariable.

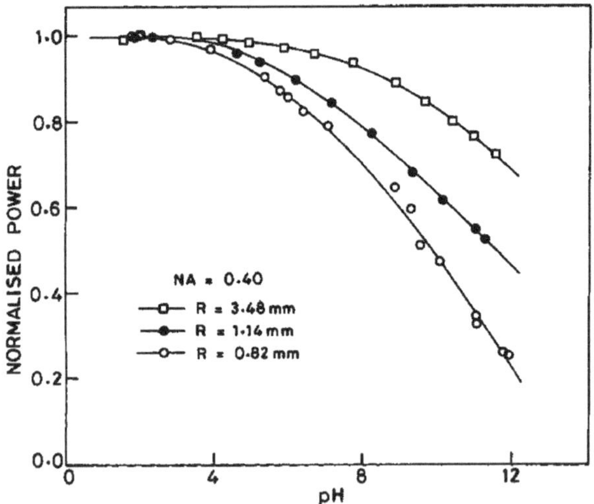

Figure 6.15: Calibration curves of the fiber optic pH sensor based on U-shaped probes of bending radii 0.82, 1.14, and 3.48 mm and numerical aperture of the fiber equal to 0.40. Reprinted from ref. 11 with permission from Elsevier.

Many fiber optic components such as couplers, polarizers, modulators, filters, amplifiers *etc* have been developed using side-polished single mode optical fibers[12-15]. Inside polished fiber, cladding is partially removed from one side of the fiber (see Fig. 5.12). If this side is brought in contact of some medium the mode field of the fiber is modified. This principle is used to sense physical and chemical parameters of a fluid. A fiber optic pH sensor is fabricated using a side polished single mode optical fiber[16]. The sensor is prepared by fixing the fiber in a groove made on one of the surfaces of a fused silica block. To make the surface sensitive to pH it is polished up to the core of the fiber and a film containing a mixture of three pH sensitive dyes is coated on it using sol-gel method. When light is launched in the fiber the evanescent field present in the film interacts with the dyes in the presence of H^+ ions. The strength of interaction depends on the pH of the fluid and accordingly the transmitted power at the other end of the fiber is changed. The main advantage of the side-polished single mode fiber is that a very small amount of sample is needed to measure the pH of the fluid. This is important when the procurement of the sample in large quantity is not possible.

To fabricate the side-polished single mode fiber sensor, first the fabrication of the groove on a fused silica/quartz block is needed. To make the groove the block is mounted on a XYZ movement and is fixed below the diamond coated wire which moves to and fro over the block with the help of a DC motor (Fig. 6.16). The top surface of the block and the wire are made exactly parallel and then the block is raised to provide appropriate pressure on the block. The uniform pressure is applied over the entire length of the wire as the radius of curvature of the groove depends on this pressure. Diamond paste and 1-2 drops of

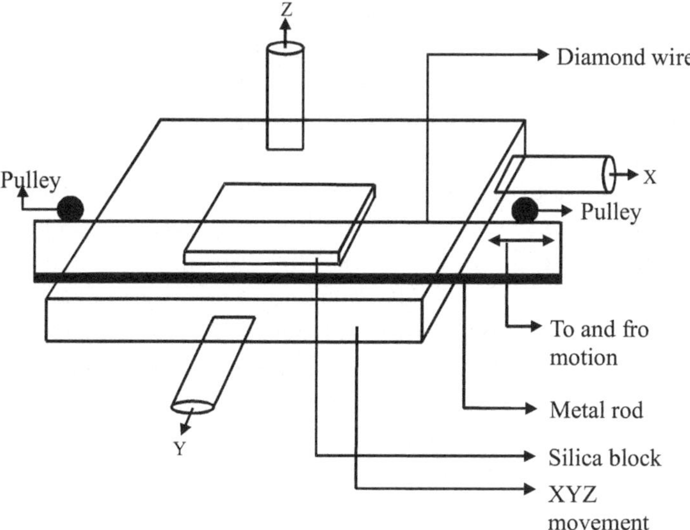

Figure 6.16: The schematic diagram for the machine used to make the groove of desired radius of curvature on a silica/quartz block. Reprinted from ref. 16 with permission from Elsevier.

glycerin oil are used to reduce the frictional heat generated due to the movement of the wire over the surface of the block. When the depth at the center of groove is found to be sufficiently enough to house the fiber (without jacket) comfortably, the deepening of the groove is stopped. The block is then cleaned in an ultrasonic bath with acetone to ensure that no glass particle is left inside the groove. The radius of curvature of the groove is used to estimate the required depth of polishing of the surface of the block. Grooves of different radii of curvature can be made on the surface of the block by applying different pressures over the block. The radius of curvature of the groove (R) is measured using a traveling microscope and the following expression

$$R^2 = \left(\frac{L}{2}\right)^2 + (R - h + d)^2,$$

where L is the length of the block, h is the depth of groove at the end of the block, and d is the depth of the groove at the center of the block.

Single mode fiber is used in the side polished half block. The plastic jacket from the fiber around the center is removed. This part of fiber is then dipped into a petridish containing acetone for few seconds to remove any trace of the remaining jacket. The groove is filled from one side with an epoxy prepared by mixing one drop of Epotec 330A (Hardner) and seven drops of 330B (Fixer). The fiber is placed gently inside the groove and the weights are placed on both sides of the block on the fiber. The block is then heated around 80°C until the color of the epoxy in the groove changes to brown. This generally takes approximately an hour. The block is then allowed to cool. Finally, the polishing of the block is carried out using polishing machine. It involves the polishing with different abrasives. First polishing was done by a mixture of abrasive (Al_2O_3) of particle size 3 μm with de-ionized water. To ensure that the cladding portion is reached, the polished region is viewed from the top with the help of a traveling microscope. The presence of an ellipse in the polished region confirms the beginning of polishing. The length of the major axis of the ellipse is measured by means of a traveling microscope. This is used to estimate the depth of the polishing already achieved and balance to be done in order to approach the core. The depth of polishing is equal to $l^2/2R$, where l is the length of the semi-major axis of the ellipse. Polishing is carried out with 3 μm particle size until 18 μm of remaining cladding is left. Next a second polishing is done by a mixture of abrasive of particle size of 1 μm and de-ionized water till approximately 8 μm of cladding is left. The final polishing is done by utilizing opaline (Cerium Oxide) with the particle size of 0.5 μm in the form of a paste in de-ionized water. After fabrication of half block the surface of the block is coated with a film of pH sensitive dyes using sol-gel method described above. In the sensor light is launched into one of the ends of the side polished fiber with the help of a suitable microscope objective. The power coming out from the other end of the side polished single mode fiber is measured by using a silicon detector along with a power meter. To characterize the probe few drops of fluids of different pH are placed at the center of the polished surface using a syringe and the corresponding output powers are measured.

6.6 Carbon Dioxide Level in Blood

Accurate and rapid measurement of the CO_2 level in blood is critical to good medical practice and patient care. To measure, blood samples are withdrawn from the patient and sent to the clinical laboratory to determine the carbon dioxide content (as part of a blood gas analysis procedure). The delays and potential errors that can be introduced because the laboratory is far from the patient and from the physician who interprets the test results, may cause therapeutic decisions to be made without adequate information. If a large number of measurements are to be made per day then withdrawing sample and taking to the laboratory each time is not practical. Further, elderly, critically ill, or infants simply cannot afford to lose any blood. Fiber optic CO_2 sensor with miniaturized probe can overcome these problems. It can be fitted into a catheter that can be inserted into blood vessels. Thus continuous, direct, minimally intrusive in vivo monitoring of the partial pressure of blood carbon dioxide can be carried out using fiber optic CO_2 sensor.

The pH sensors described above can also be used as CO_2 sensor if the probe is covered with the bicarbonate solution which is separated from the sample by a gas permeable membrane. Carbon dioxide from the sample diffuses through the membrane until an equilibrium is reached between the partial pressure of CO_2 in the sample solution and the CO_2 in the probe. The CO_2 that diffused across the membrane changes the pH of the bicarbonate solution. The change in pH is measured by the pH sensor. Increase in pCO_2 decreases the pH of the solution.

6.7 Carbon Monoxide Sensor

Carbon monoxide is a poisonous and explosive gas. It is produced if the combustion of carbon compounds or carbonaceous materials is incomplete. It is also a compound of nearly all types of gases produced on heating, exhaust gases of automobiles, flue gases produced by home-wood, coal or oil. Its detection is very important for industrial as well as non-industrial uses. Carbon monoxide is frequently detected with an electrical-based detector and there is always a possibility of spark with the electrical-based detector. In such an eventuality, if the concentration of carbon monoxide is above the explosion limit then the electrical spark can ignite an explosion. To avoid the risk of explosion carbon monoxide sensors utilizing optical fibers have been developed. Although there are number of fiber optic carbon monoxide sensors however here we shall describe a reagent-mediated fiber optic carbon monoxide sensor based on a porous polymer optical fiber[17].

Palladium chloride, a colorimetric indicator, is used to sense the carbon monoxide. On reacting with carbon monoxide, palladium chloride changes its colour from yellow to black. The design of the probe is similar to that shown in Fig. 6.10. The dye is trapped within the polymer matrix during the preparation of the porous polymer optical fiber. One of the disadvantages of the indicator is that it is degraded when it is exposed to the atmosphere. However, the hydrophobic property of the polymer and a kind of binding between the palladium chloride with the polymer make the indicator more stable to the atmosphere. Further, exposure to hydrogen and petrol vapour does not affect the indicator. However, hydrogen sulphide exhibits a significant effect on the absorption spectrum of the indicator. The effect is in the spectral range shorter than 600 nm. Thus the interference of hydrogen sulphide to the detection of carbon monoxide is eliminated by selecting the wavelength of the light greater than 600 nm.

In the sensor, light from a high intensity quartz halogen lamp with monochromator (630 nm) is injected into the porous polymer sensing fiber through plastic optical fibers spliced to its both the ends (Fig.6.17).

A 1 cm length of the porous polymer fiber is used for the probe. Teflon capillaries of internal diameter slightly smaller than that of the fiber are used to connect the fibers. The probe is placed in a gas chamber. A gas flow system controls the introduction of carbon monoxide in the chamber. The output signal is detected by a silicon photodiode. When carbon monoxide comes in contact of the probe it diffuses into the porous polymer fiber. The gas then reacts with the indicator to create the metallic palladium black particles. These particles cause the light scattering loss and a decrease in observed light intensity. In the probe equilibrium between the carbon monoxide and the indicator does not take place. That means the palladium chloride will keep on reacting with the carbon monoxide until all of the palladium chloride is consumed. Thus the stabilized signal cannot be reached before whole of the indicator is consumed. In such a situation the sensing fiber probe turns black. Thus the sensor is best suited for alarm systems that detect the leakage of the carbon monoxide. Since it can not be reused it serves as a "fuse" which can be easily changed by plugging-in and unplugging from the regular optical fiber cables. To reduce the cost of the sensor LED of wavelength greater than 600 nm can be used.

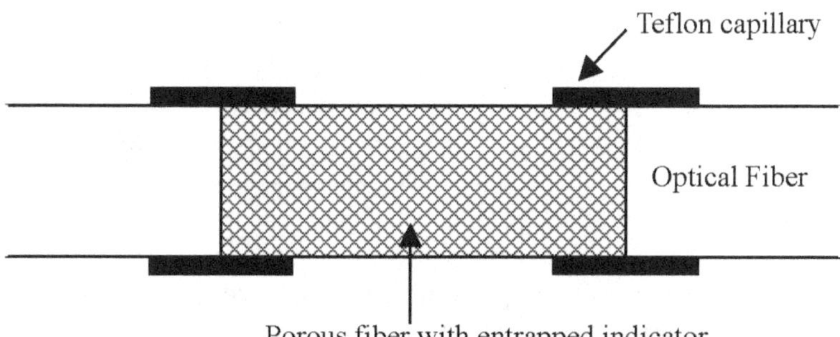

Figure 6.17: Carbon monoxide probe based on porous polymer optical fiber with entrapped indicator.

6.8 Hydrogen Sensor

Hydrogen is another flammable gas. Similar to carbon monoxide its detection using optical fibers is advantageous. This is because in the probe no wires are used that could result in a spark igniting the gas. The hydrogen sensor described below is based on the changes in reflectivity of thin palladium film deposited on the end face of an optical fiber when exposed to hydrogen gas[18]. A schematic of the sensor is shown in Fig.6.18. Light from an LED operating at 860 nm wavelength is injected into one of the arms of an optical fiber coupler. The light passes through the coupler to the sensing fiber end where the palladium film is deposited. Part of the light reflected from the fiber end passes through the coupler to the photodetector where it is detected. When hydrogen gas comes in contact of the palladium film changes in the optical properties of Pd film occurs. The change causes the reduction in the reflectivity of the film. Thus the reflected power decreases as the amount of hydrogen gas increases. The size of the response depends on the H_2 concentration. A very thin (10 nm) Pd film shows large changes in reflectivity upon exposure to hydrogen gas.

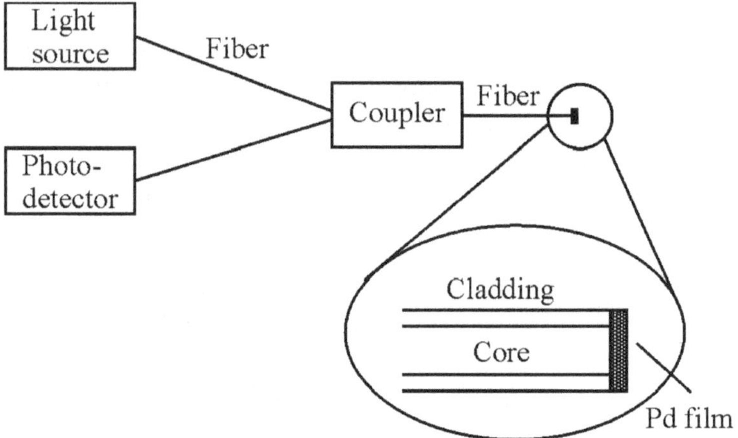

Figure 6.18: Schematic of an optical fiber hydrogen sensor. Reprinted from ref. 18 with permission from Elsevier.

Another hydrogen sensor consists of a cladded multimode tapered fiber coated with a thin palladium layer[19]. The evanescent field in the tapered region interact with the palladium film. The presence of hydrogen gas converts the palladium layer into a palladium hydride film. As a result, the evanescent field is absorbed by the film and the guided power suffers attenuation which can be detected with a simple detection scheme. The sensor's performance depends on the taper ratio. The response time of the sensor is approximately 30 ns and can monitor hydrogen concentration below 4%.

6.9 Aluminium Sensor

Aluminium is widely known to be associated with some diseases such as bone softening. This occurs mainly due to the presence of Al(III) in drinking water. Thus an Al(III) sensor which can operate remotely and can monitor the water quality continuously is required. These requirements can be fulfilled using optical fiber for sensing. In this section we describe a fiber optic Al(III) sensor based on diffuse reflectance spectrophotometry[20]. A reagent eriochrome cyanine R (ECR) is immobilized on XAD-2 (styrene-divinylbenzene cross-linked copolymer). It is encapsulated at the end of optical fiber using nylon mesh (Fig. 6.19). When Al(III) comes in contact of ECR it changes the colour of ECR from red to dark blue. This results in the decrease in the reflectance of immobilized ECR. The reflectance depends on Al(III) concentration. The maximum reflectance change occurs at 643 nm wavelength. The measurements are expressed as a relative reflectance which is defined as the difference between the reflectance of the Al(III) complex and that of the immobilized ECR alone, both recorded at 643 nm.

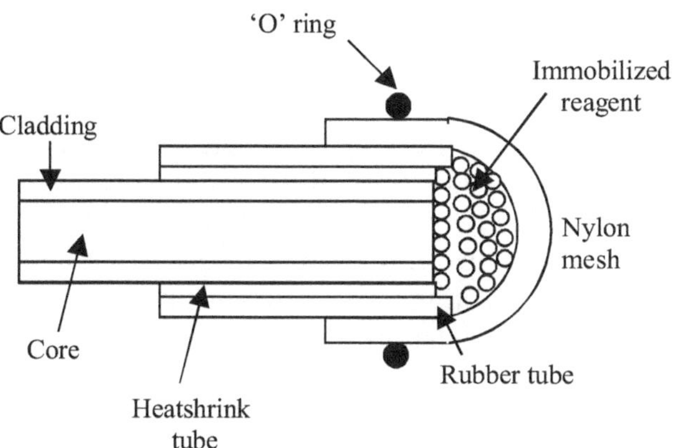

Figure 6.19: Fiber optic Al(III) probe. Reprinted with permission. Reprinted from ref. 20 with permission from Elsevier.

In the apparatus used to measure reflectance a bifurcated optical fiber is used. Light from a quartz-halogen lamp is launched in one of the two branches of the bifurcated fiber. It reaches the probe which contains XAD-2 immobilized with ECR. The light reflected back is coupled to the fiber. It is then guided by the second branch of the bifurcated fiber to a detection system equipped with a monochromator

(643 nm). A linear relationship is obtained between the relative reflectance and the logarithm of Al(III) concentration in the range of $1.3x10^{-5}$ M to $4.0x10^{-4}$ M with the limit of detection of $1.0x10^{-5}$ M. The response of the probe is affected by the pH of the solution. Thus the measurements are carried out at fixed pH. The size of the signal increases with the increase in the ECR concentration used for immobilization. After Al(III) measurement, the probe is dipped in a regenerating solution to remove the Al(III) ions from the probe. EDTA, tartrate, fluoride, oxalate or phosphate can be used as regenerating agent. It has been reported that Be^{2+} interferes in Al(III) determination.

6.10 Ammonia Sensor

Several different approaches toward the development of fiber-optic ammonia sensors have been reported in the literature. Here we shall describe an ammonia sensor based on chromophoric indicator dye. The sensor is fabricated by trapping the internal indicator solution inside a short segment of gas-permeable tube. The optical fibers are used to supply incident radiation and to collect the light that traverses through the internal indicator solution. The indicator solution is composed of ammonium chloride, sodium chloride and a suitable chromophoric pH indicator. Sodium chloride is added to match the osmolarities between the sample and internal solutions across the wall of the tube. Figure 6.20 shows a schematic diagram of the various phases and dynamic equilibria involved in the response of the ammonia probe. Ammonia from the sample solution diffuses across the gas-permeable membrane, enters the internal solution and reacts with the indicator dye. The diffusion continues until the ammonia partial pressure is equal on both sides of the membrane and a steady state ammonia concentration is established in the indicator solution. Variation of the ammonia concentration in the internal solution causes a change in the pH of this solution which alters the relative concentration ratio of two forms of the pH indicator dye. In the ammonia sensor described below[21], bromothymol blue has been used as the pH indicator dye. As the ammonia concentration in sample increases, the amount of ammonia in the internal solution also increases which results in a higher pH of internal solution and more of the indicator dye in the non-protonated form. The non-protonated form of the indicator is detected by absorption measurement.

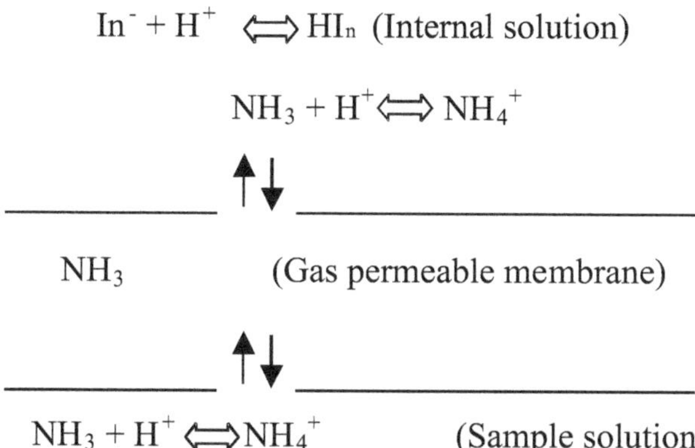

Figure 6.20: Various phases and chemical equilibria involved in ammonia sensor.

The schematic of the fiber optic ammonia probe is shown in Fig.6.21. The fiber bundles are used to supply incident radiation and to collect light that traverses through the internal solution. The absorption of light in the solution increases as the concentration of ammonia increases. The sensor's response time is within 13-16 min for 200 nM level of ammonia with the detection limit is around 100 nM.

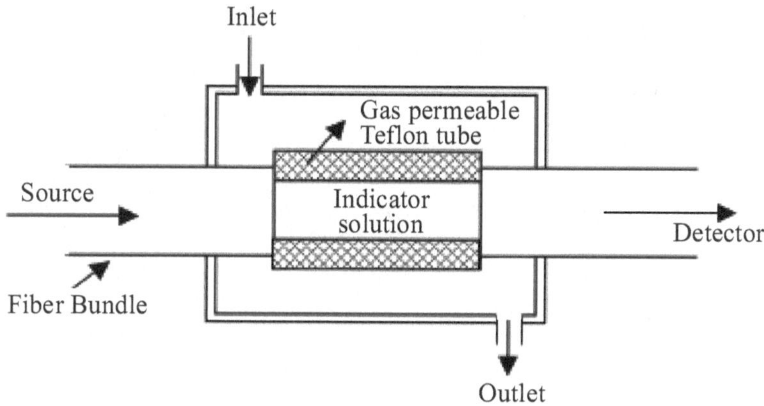

Figure 6.21: Fiber optic ammonia probe.

6.11 Chloride Sensor

Chloride in the form of salt water is a major contaminant of ground water, percolating through landfill liners and causing corrosion of steel. The tests used to determine either steel corrosion or chloride contamination of ground water are time-consuming and not suited for long-term in situ monitoring. In the conventional methods the water samples are collected from the wells and transported to a laboratory. In the case of concrete structures, tests for salt intrusion near the steel reinforcement require the grinding and removal of concrete samples that are transported to the laboratory and tested. This testing is very laborious and requires numerous samples to pinpoint a trouble spot. Optical fiber sensors can be used to detect chloride concentration in situ. These can be embedded near or on steel rods or at the bottom of a landfill. These can detect the presence of salt as it disperses through the concrete or protective landfill layers. This procedure eliminates sampling delays, a significant amount of labour and chance of error in measurements. The other advantages with optical fibers are their immunity to corrosion and chemical stability of the plastics and glasses these are made of.

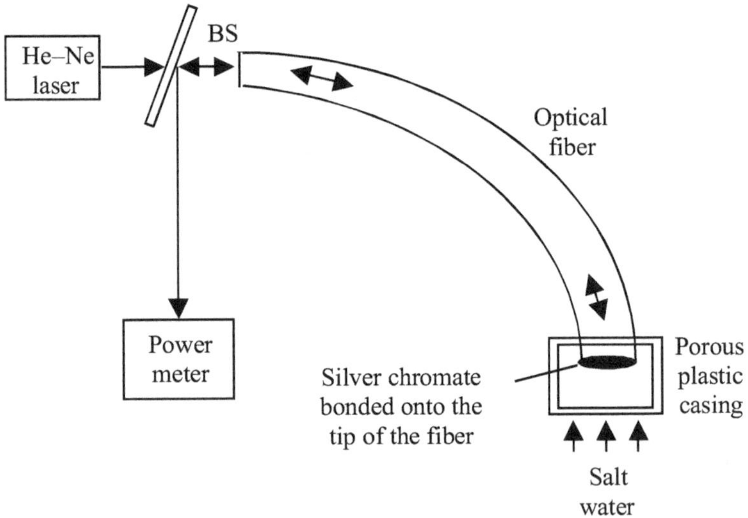

Figure 6.22: Chloride sensor based on reflective concept[22]. Reprinted with permission.

An extrinsic type reagent-mediated fiber optic sensor for the detection of chloride ions[22] is shown in Fig.6.22. The silver chromate (Ag_2CrO_4) powder is bonded to one of the ends of a plastic optical fiber of 1 mm core diameter using a thin layer of translucent waterproof glue. Light from a He-Ne laser is coupled into the other end of the fiber. The light reflected from the sensing end is separated from the input light using an optical beam splitter and detected using a power meter. When the probe is introduced in a tank containing salt water, the chloride ions diffuse to the silver chromate and changes its colour from brown red to white. This occurs due to the formation of silver chloride. The change in colour increases the reflectivity of the probe and hence the increase in the light received by the detector. The response time of the sensor is about 5 min.

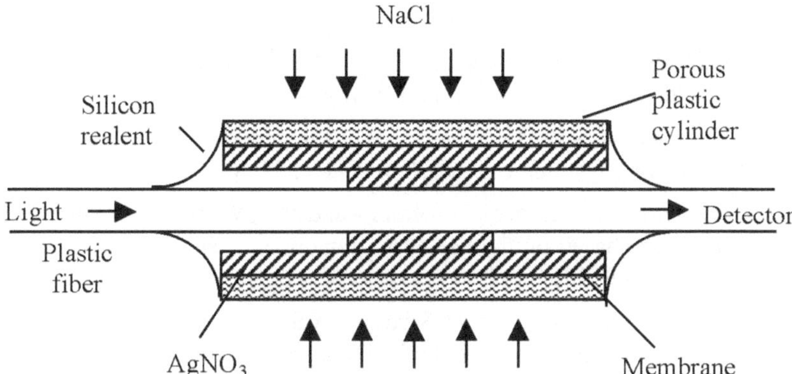

Figure 6.23: Reagent-mediated chloride sensor utilizing evanescent field absorption[22].

A fiber optic salt sensor based on evanescent wave absorption[22] is shown in Fig.6.23. The fiber with glass core and acrylic plastic cladding is used for the sensor. Few cm length of the cladding is removed from the middle portion of the fiber. It is then encapsulated by a silver nitrate solution, a membrane that only allows passage of chloride ions and a porous plastic cylinder. The porous plastic cylinder filters out soil particles and allows leachate into the sensing region. Light is launched from one end of the fiber and is detected at the other end. If the sensor is immersed into the salt water solution, chloride ions penetrate through the cylinder and the membrane and react with silver nitrate ($AgNO_3$) to form a white silver chloride (AgCl) precipitate. The white precipitate absorbs some of the evanescent field present in the sensing region. This results in the decrease in the light received by the detector. The response time of the sensor is about 1 min.

References

1. Q. Zhou, M.R. Shahriari, D. Kritz and G.H. Sigel, Jr. (1988) Porous fiber-optic sensor for high-sensitivity humidity measurements. *Anal. Chem.* **60**, 2317-2320.
2. J.Y. Ding, M.R. Shahriari and G.H. Sigel, Jr. (1991) Development of high temperature humidity sensors based on porous optical fibers. *Int. J. Optoelectron.* **6**, 385-393.
3. S. Muto, A. Fukasawa, T. Ogawa, M. Morisawa and H. Ito (1990) Optical detection of moisture in air and in soil using dye doped plastic fibers. *Jap. J. Appl. Phys.* **29**, L1023-L1025.
4. B.D. Gupta and Ratnanjali (2001) A novel probe for a fiber optic humidity sensor. *Sensors and Actuators* B **80**, 132-135.
5. J.I. Peterson, S.R. Goldstein and R.V. Fitzgerald (1980) Fiber optic pH probe for physiological use. *Anal. Chem.* **52**, 864-869
6. G.F. Kirkbright, R. Narayanaswamy and N.A. Welti (1984) Fibre-optic pH probe based on the use of an immobilised colorimetric indicator. *Analyst* **109**, 1025-1028.
7. J. Liu, M.H. Shahriari, G.H. Sigel and A.S. Goldman (1993) Fibre optic pH sensor based on a chemically modified porous polymer film immobilized with colorimetric indicator. *Int. J. Optoelectron.* **8**, 193-201.
8. J.Y. Ding, M.R. Shahriari and G.H.Sigel (1991) Fibre optic pH sensors prepared by sol-gel immobilisation technique. *Electron. Lett.* **27**, 1560-1562.
9. B.D. Gupta and D.K. Sharma (1997) Evanescent wave absorption based fiber optic pH sensor prepared by dye doped sol-gel immobilization technique. *Optics Communication* **140**, 32-35.
10. B.D. Gupta and S. Sharma (1998) A long range fiber optic pH sensor prepared by dye doped sol-gel immobilization technique. *Optics Communications* **154**, 282-284.
11. B.D. Gupta, and N.K. Sharma (2002) Fabrication and characterization of U-shaped fiber optic pH probes. *Sensors and Actuators B* **82**, 89-93.
12. R.A. Bergh, G. Kotler and H.J. Shaw (1980) Single mode fiber optic directional coupler. *Electron. Lett.* **16**, 260-261.
13. D. Gruchmann, K. Petermann, L. Staudigel and E. Weidel (1983) Fiber optic polarizers with high extinction ratio. *Proc. 9th Eur. Conf. Optical Commun.,* Geneva, Switzerland, North Holland, Amsterdam, p. 305.

14. W.V. Sorin, K.P. Jackson and H.J. Shaw (1983) Evanescent amplification in a single mode optical fiber. *Electron. Lett.* **19**, 820-821.

15. C.A. Millar, M. Brierley and S. Mallinson (1987) Exposed-core single mode fiber channel-dropping filter using a high index overlay waveguide. *Opt. Lett.* **12**, 284.

16. N.K. Sharma and B.D. Gupta (2003) Fabrication and characterization of pH sensor based on side polished single mode optical fiber. *Optics Communications* **216**, 299-303.

17. Q. Zhou and G.H. Sigel Jr. (1989) Detection of carbon monoxide with a porous polymer optical fiber. *Int. J. Optoelectron.* **4**, 415-423.

18. M.A. Butler (1994) Micromirror optical-fiber hydrogen sensor. *Sens. Act.* B **22**, 155-163.

19. J. Villatoro, D. Luna-Moreno and D. Monzon-Hernandez (2005) Optical fiber hydrogen sensor for concentrations below the lower explosive limit. *Sensors and Actuators* B **10**, 23-27.

20. M. Ahmad and R. Narayanaswamy (1994) Fiber optic reflectance sensor for the determination of aluminium (III) in aqueous environment. Anal. Chim. Acta **291**, 255-260.

21. S. Kar and M.A. Arnold (1994) Cylindrical sensor geometry for absorbance-based fiber-optic ammonia sensors. *Talanta* **41**, 1051-1058.

22. P. Cosentino, B. Grossman, C. Shieh, S. Doi, H. Xi and P. Erbland (1995) Fiber-optic chloride sensor development. *J. Geotechn. Engg.* **121**, 610-617.

7

Wavelength Modulated Sensors

7.1 Introduction

There are numerous wavelength dependent phenomena that may be used to modulate the wavelength of incident light source. These are wavelength dependent absorption, luminescence, dispersion, interference and scattering. Wavelength modulated sensors use a broadband source, a wavelength modulator or measurand (*i.e.* analyte), a form of spectrometer (using prisms, gratings or interference filters), a detector (single photodiode or array) and signal processing unit. Silicon photodiode is cheap and can be used for detection but it limits the detection wavelengths to the region 400 to 1050 nm. The sensitivity or the resolution of the wavelength modulated sensor depends on the characteristics of the detector, the inherent resolution of the spectrometer, mechanical stability of the modulation unit and the spectrometer, and the capability of the signal processing unit. However, wavelength modulated sensors do not require any form of referencing like intensity modulated sensors. In intensity modulated sensors, loss in intensity occurs due to connectors and splices, microbending, mechanical creep and misalignment of light sources and detectors due to environmental effect. Further the fluctuations in light source can affect the performance of the sensor. Thus, for intensity modulated sensors one needs referencing to cancel out these losses. In some intensity modulated sensors two wavelengths are used. One wavelength is used to cancel out all of the losses while the second is used to sense the measurand.

Wavelength detection is relatively easy if the spectral changes result from changes in chemical indicator dyes, fluorescence, phosphorescence, or black-body radiation. In these cases the incident and emitted wavelengths are far apart and therefore detection is easy. If the spectral change occurs due to mechanical movement, then the wavelength filtering mechanisms must have sufficient resolvable spectral points. The reasons for using fluorescence in the fiber optic sensors are well known. Fluorescence is intrinsically more sensitive than absorbance and is more flexible in that a great variety of analytes and physical parameters are known to change the emission of particular fluorophores. The fluorescence based sensors utilize the change in the intensity, lifetime or colour of fluorescence as the affected parameter by the measurand. That means, any interpretable observation of fluorescence can be made the basis of a sensor. In this chapter we shall first describe luminescence phenomenon and then various physical and chemical fiber optic sensors based on wavelength modulation and fluorescence phenomenon.

7.2 Luminescence

When light from a broadband source is incident on a transparent layer of solid, liquid or gas, certain wavelengths of the light source may be absorbed by the atoms and molecules. The absorption results in the excitation of the atoms or molecules into a higher energy state. The excited species is short-lived and releases its surplus energy in several ways. If the energy is released in the form of light then, depending on whether the excited state is singlet or triplet, the emission is called fluorescence or phosphorescence. In fluorescence, an atom or molecule makes an allowed transition from a higher to a lower energy state in a short time (10^{-9} to 10^{-3} sec). It occurs at a longer wavelength than the exciting wavelength and has a uniform spatial distribution. If the excited state is produced by some other means then the process is called by the different name. For example, if the excited state is produced by a chemical reaction then the process of emission of light is called chemiluminescence. In the case of photoexcitation it is photoluminescence while in the case of electrical excitation it is called electroluminescence.

7.2.1 Fluorescence

The characteristics of fluorescence such as intensity, life time and energy transfer can be used to sense various chemical and biological analytes. Below we describe these characteristics.

Intensity

There are three parameters that characterize the fluorescence. These are excitation wavelength, emission wavelength and emission intensity. The emission wavelength is always greater than the excitation wavelength while the emission intensity depends on the concentration of the luminescent species. The fluorescence intensity, I, and analyte concentration, c, are related according to the following equation

$$I = 2.3kI_0\phi_f \varepsilon cL\left[1 - \frac{2.3\varepsilon cl}{2!} + \frac{(2.3\varepsilon cl)^2}{3!} + --- \right] \quad (7.1)$$

where I_o is the intensity of the exciting light, ε is the molar absorption coefficient at the wavelength of excitation, ϕ_f is the fluorescence quantum yield, L is the optical depth and k is a factor that depends on the geometrical arrangement of the instrument. For weakly absorbing solutions ($\varepsilon cL < 0.05$), above equation simplifies to

$$I = 2.3kI_0\phi_f \varepsilon cL \quad (7.2)$$

A large number of sensors have been developed on the basis of fluorescence intensity. There are various factors that can affect the fluorescence intensity and hence one may not get accurate measurement. This can be reduced significantly if a constant concentration of luminescent species (L) is used with an analyte (A) that quenches the fluorescence. The mechanism can be depicted by

$$L + h\nu \xrightarrow[k_e]{\text{excitation}} L^* \xrightarrow[k_l]{\text{luminescence}} L + h\nu'$$

$$+$$

$$A$$

$$\downarrow k_q \text{ quenching by analyte (A)}$$

$$L + A^*$$

where k_e, k_l and k_q are the rate constants (in different units) of excitation, fluorescence and quenching processes respectively. The rate equation for the above mechanism can be written as

$$\frac{d[L^*]}{dt} = k_e I_o [L] - k_l [L^*] - k_q [L^*][A] \quad (7.3)$$

At equilibrium under steady illumination

$$[L^*](k_1 + k_q[A]) = k_e I_o [L] \tag{7.4}$$

or,

$$k_1[L^*] = \frac{k_e I_o [L]}{1 + K_{SV}[A]} \tag{7.5}$$

where

$$K_{SV} = \frac{k_q}{k_1}$$

When [A] = 0 *i.e.* in the absence of quencher,

$$k_1[L^*] = k_e I_o [L] = I_a \tag{7.6}$$

In the presence of quencher, the fluorescence intensity is

$$I_1 = \frac{I_a}{1 + K_{SV}[A]}$$

or,

$$\frac{I_a}{I_1} = 1 + K_{SV}[A] \tag{7.7}$$

Equation (7.7) is called the Stern-Volmer equation and K_{SV} is called the Stern-Volmer quenching constant. Thus by knowing the ratio of the two intensities, the concentration of the analyte can be determined.

Life Time

Another parameter that has been exploited for sensing is the fluorescence life-time. It is defined as the average time a molecule remains in the excited state. The typical lifetime of a fluorophore ranges from 1-100 nsec while the phosphorescence lifetimes are much longer typically 1-1000 μsec. This is because the triplet states are long-lived. There are two methods that have been widely used for the determination of life time. These are pulse method and the phase modulation method. In the former, the sample is excited with a brief pulse of light and the time dependent decay of fluorescence intensity is measured. In the phase modulation method, the sample is excited with sinusoidally modulated light and phase shift between the sine function of fluorescence to that of exciting light is determined.

Sensors based on life-time measurements are useful in optical sensing if the analyte affects the life-time of a fluorescent species. The relation between life-times in the absence (τ_o) and presence (τ) of a quencher is given by the equation similar to Eq. (7.7)

$$\frac{\tau_o}{\tau} = 1 + K_{SV}[A] \tag{7.8}$$

As can be seen there is a linear relationship between the analyte or quencher concentration and $\left(\frac{I_o}{I} - 1\right)$ or $\left(\frac{\tau_o}{\tau} - 1\right)$. Sensors based on life-time measurements require more sophisticated optoelectronic instrumentation. However, the life-time based sensing has some advantages over fluorescence intensity-based sensing.

Energy Transfer

Fluorescence energy transfer is the transfer of excited-state energy from a donor (D) to an acceptor (A). This transfer occurs without the appearance of a photon. It is primarily the result of dipole-dipole interactions between the donor and acceptor as shown below

$$D \xrightarrow{hv} D^*$$

$$D^* + A \longrightarrow D + A^*$$

$$A^* \longrightarrow A + hv'$$

The fluorescence intensity from the second species, A, is proportional to $k_{et}[D^*][A]$ where k_{et} is the rate of energy transfer between the two species. Energy transfer process depends strongly on (i) the amount of overlap between the emission spectrum of the donor and the absorption spectrum of the acceptor, (ii) the distance between the donor and acceptor, and (iii) the relative orientation between the donor and acceptor transition dipoles. It is the dependence on the overlap of the spectra and the distance that have been exploited in sensors. Typical critical distances over which energy can be transferred are 0.5 to 10 nm.

7.2.2 *Phosphorescence*

Few sensors have been developed based on phosphorescence. The fundamental relations between analyte concentration and phosphorescence intensity and life-time are the same as in fluorescence. Because of longer life-times of triplet state, much smaller quencher concentrations are sufficient to cause the same quenching efficiency (intensity or life-time reduction) as in fluorescence. It means that phosphorescence quenching is more sensitive.

7.2.3 *Chemiluminescence*

If a chemical reaction is accompanied by the emission of light then the phenomenon is called chemiluminescence. It can be depicted as

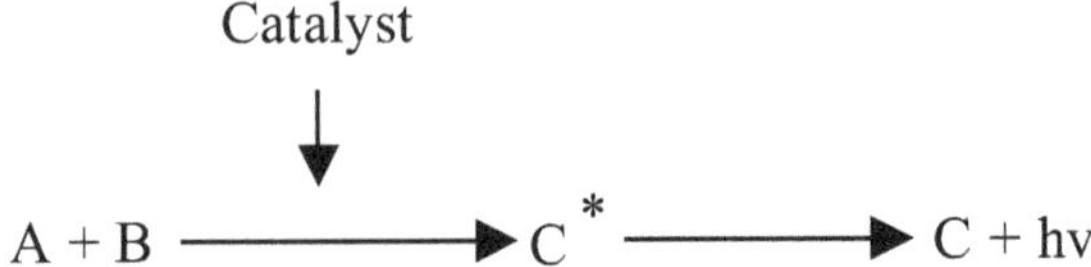

Any of the components of this reaction, including the catalyst can be measured. Since the signal is transient, the measurement of chemiluminescent intensity is time-dependent. The signal is often recorded at a specific time after interaction of reagent and analyte or by integration of light over a fixed time. The main advantages of chemiluminescence methods are low detection limits, wide dynamic ranges and the fact that there is no need for light source with its power requirements since the chemically sensitive material itself is the emitter. Thus the instrumentation for this method is quite simple.

Now we shall describe number of sensors utilizing wavelength modulation.

7.3 Displacement sensor

Figure 7.1 shows a simple prism based lateral displacement sensor. White light is launched in an optical fiber using a suitable optics. Light exiting from the fiber is collimated with the help of a lens. The collimated beam then falls on a prism that causes its dispersion. A second fiber (also called output fiber) is placed at the focal plane of the second lens on the other side of the prism. The output fiber collects the light from a small area of the spectrum. If the prism is rotated about the axis normal to the plane of the paper the wavelength reaching at the input face of the output fiber changes. This wavelength is analyzed at the other end of the fiber. By measuring the wavelength the angular displacement of the prism can be determined. Alternatively, if the prism is kept fixed, the lateral displacement of the output fiber may be sensed.

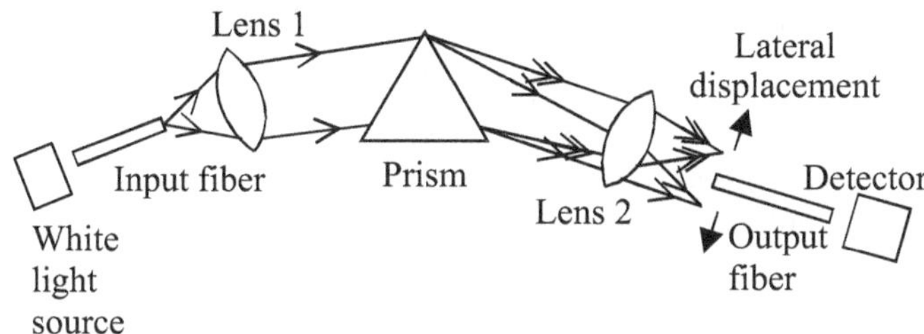

Figure 7.1: Prism based lateral displacement sensor.

The displacement can also be sensed if a reflection grating is used in place of a prism. The experimental set up is shown in Fig. 7.2. Light from a broadband source is launched into an input fiber. The light exiting from this fiber is then collimated using a lens. The collimated beam is allowed to fall on a reflection grating which diffracts the beam. The output fiber collects a part of the diffracted beam (spectrum). If the grating is rotated the wavelength of light received by the output fiber changes. By measuring the wavelength the angular rotation of the grating can be determined.

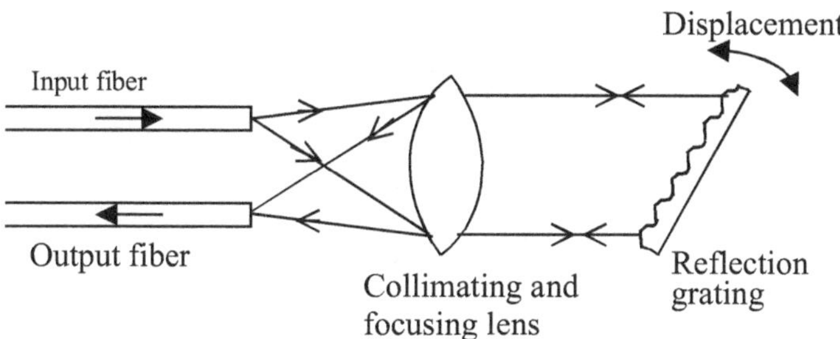

Figure 7.2: Grating based displacement sensor.

Both the sensors described above are highly immune to source and fiber transmission variations. Their success depends on the design of the wavelength decoder and on the availability of a broadband source capable of meeting the cost and environmental constraints.

7.4 Temperature sensors

The measurement and control of temperature is of great importance in every area of scientific and technological research. The important applications are in industrial processes, environmental regulatory systems and biomedical sciences. Almost all chemical processes and reactions are temperature dependent. In chemical plant, temperature is the only indication of the process which is taking place. In some cases, loss of control of temperature can result in catastropic plant failure with attendant damage and possible loss of life. There are many other areas of industry where temperature measurement is essential. These applications include steam raising and electricity generation, plastics and glass manufacture and moulding, metallurgical industries, oil refineries, coal mines, jet engines, nuclear reactors, milk and dairy products and many other aspects of food industries. In electrical power stations the monitoring and control of temperature is required in high voltage power machines such as transformers, generators and motors. The conventional temperature monitoring devices such as thermocouple cannot be used inside such machines because of electromagnetic noises or short-circuiting. For such an application, fiber optic temperature sensors are most advantageous. Fiber optic temperature sensors are also useful in many industrial processes where treatment of aqueous solutions under a strong electromagnetic field is involved such as in the caustic industry. In biomedical field, thermometers are required for continuous monitoring of temperature in RF or microwave hyperthermia for treatment of some cancers, in phototherapy of malignant tumours and in other medical procedures such as laser surgery, electrosurgery, cryosurgery and in monitoring blood flow. The cardiac output measurements by the conventional thermodilution technique also require monitoring of temperature. In the treatment of cancer, the production of localized and controlled hyperthermia by electromagnetic energy poses a difficult temperature measurement problem. The conventional temperature sensors, such as thermistors or thermocouples, require metallic components and connecting wires, which significantly perturb the incident electromagnetic fields and may cause undesirable localized heating. To overcome these problems, fiber optic temperature sensors have also been proposed for biomedical applications. The main advantage of these sensors in biomedical field is that they can be easily introduced in the catheters or hypodermic needles and can be used for highly localized measurements. Further, these are chemically safe, do not require electrical connections and are flexible and light in weight. These sensors can also measure temperature in otherwise inaccessible or difficult circumstances, such as those bathed in RF radiation, or other noisy electromagnetic environments. In this section we shall describe wavelength modulated fiber optic temperature sensors based on different phenomenon.

7.4.1 Black-body radiation

One of the most direct measurements involving wavelength is the black-body based temperature sensor. In this sensor no light source is needed to launch light in the fiber. A black body cavity that emits broadband radiation from the inner surfaces of the cavity is placed at the end of an optical fiber as shown in Fig. 7.3. The light emitted by the body is collected by the fiber and is guided up to its other end. The light exiting from this end of the fiber is collimated using a lens and then splitted into two parts by a beam splitter. The detectors in combination with narrow-band filters examine the received spectrum and hence the temperature of the black-body. As the temperature of the black-body increases the spectrum of the source shifts towards the lower wavelength region. This changes the intensities of two beams. The ratio of the intensities of the two beams will give the temperature of the black-body.

This type of sensor has been successfully commercialized. The sensor operates in the temperature range 500°C to 2000°C with a very high accuracy at higher temperatures. The sensor's resolution is 7.5×10^{-6} °C.

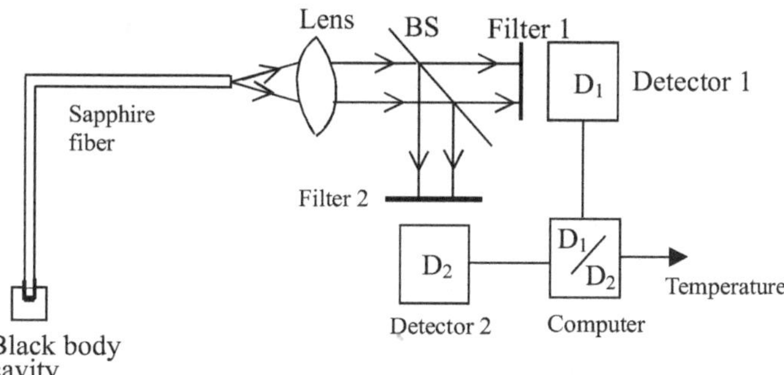

Figure 7.3: Fiber optic temperature sensor utilizing black-body.

7.4.2 Semiconductor band edge

The band-gap energy of most of the semiconductors decreases almost linearly with increasing temperature (T) near room temperature. Hence the wavelength $\lambda_g(T)$ corresponding to their fundamental optical absorption edge shifts towards longer wavelength with temperature. When the spectrum of the light source overlaps with both sides of the optical absorption edge, a shift of this edge induced by a change in the temperature modifies the fraction of the light absorbed by the semiconductor and therefore changes the intensity of the transmitted light. For example, at 25°C and atmospheric pressure the starting point of optical absorption edge of GaAs corresponds to 867 nm wavelength. This implies that GaAs is transparent to light of wavelengths greater than this value. When temperature increases the absorption band edge will shift towards the longer wavelengths resulting in the absorption of a larger part of the spectrum of the source as shown in Fig. 7.4. For a wide range of temperature, the source should be broadband. If more accurate measurements are needed then a laser diode or gas laser should be used.

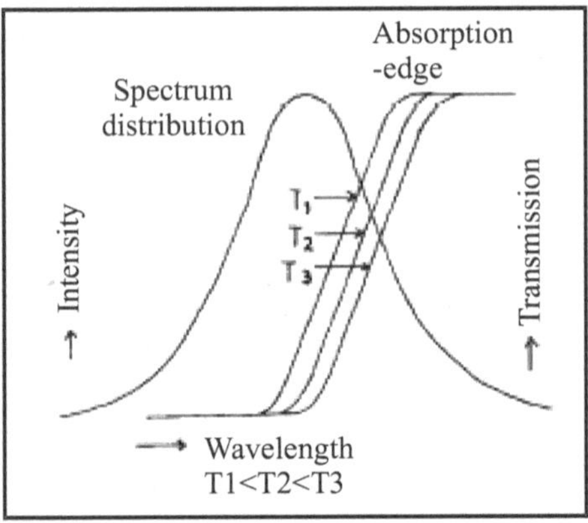

Figure 7.4: Temperature-induced shift of absorption edge of GaAs with respect to source spectrum. Reprinted with permission[1]. À© [1982] IEEE.

Based on this principle a temperature sensor is developed as shown in Fig. 7.5. Light from a broadband source is launched into one of the two fibers. It then passes through a GaAs prism and then to the second fiber after total internal reflection. The second fiber is connected to a detector. For photon energies less than the band gap energy (λ>900nm) light is not significantly attenuated in the GaAs prism while the shorter wavelengths are absorbed in the prism. Since the band gap energy is temperature dependent the absorption profile of the probe will also be temperature dependent. The band edge for the GaAs moves about 0.5 nm/°C. This implies a band-edge wavelength accuracy requirement of the order of 1 A° for a 0.1 °C sensor.

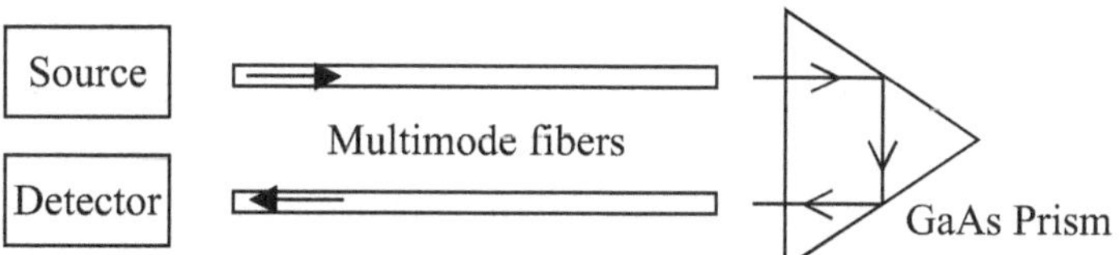

Figure 7.5: Fiber optic temperature sensor based on temperature dependent band edge shift in GaAs semiconductor.

Another temperature sensor[1,2] based on the same principle and the same semiconductor is shown in Fig. 7.6. Two different LEDs are used in the design to overcome the error caused by stray losses, transmission loss of fibers and optical connector loss. Two pulses of wavelengths λ_1 (0.87 μm) and λ_2 (1.27 μm) are guided by the optical fiber from both AlGaAs-LED and InGaAsP-LED to a germanium avalanche photodiode via a 0.2 mm thick GaAs chip. The temperature of the chip affects the transmission of the light of wavelength λ_1 while almost all of the light of wavelength λ_2 passes through the sensor independent of temperature. The ratio of the output signals of λ_1 and λ_2 gives the temperature of the probe. The accuracy of the sensor is about ± 0.5°C in the temperature range from -30 to 300°C.

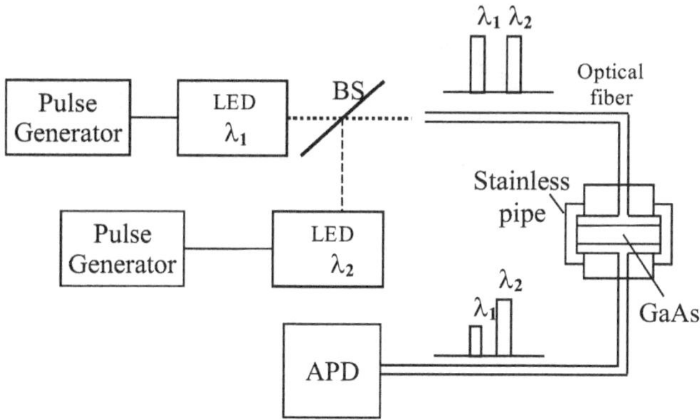

Figure 7.6: Temperature sensor utilizing GaAs crystal and two LEDs to overcome the losses.

7.4.3 Fluorescence emission

Several types of photoluminescent methods have also been used in fiber optic temperature sensors. One of the methods is based on the temperature dependent fluorescent emission from phosphors[3]. When phosphors are excited by ultraviolet light, they re-emit lower energy fluorescent light. Certain phosphors emit several lines each having different temperature dependence. The ratio of the intensity of two emission lines is independent of the excitation condition and the variations in parasitic losses but depends on the temperature. Figure 7.7(a) shows a fluorescent temperature sensor based on the detection of intensities at two wavelengths of the spectrum. The phosphor material used is gadolinium oxysulfide activated by a small amount of europium. The emission spectra of the material is shown in Fig. 7.7(b) along with the temperature dependences of two selected f_1 and f_2 lines and their intensity ratio (f_1/f_2). It can be seen that as the temperature increases the ratio decreases almost linearly. The probe consists of a 400 μm plastic clad fiber with a 200 μgm of the phosphor attached at the fiber end. Light from an ultraviolet source is launched into the fiber to excite the phosphor. The light emitted returns to the detection system through the same fiber. The returned light is divided into two beams using a beam splitter and pass through the two filters corresponding to f_1 and f_2 lines. Their intensities are detected by means of separate detectors. The ratio of the two intensities then gives the temperature of the probe. The sensor operates in the range -50°C to 250°C with an accuracy of 0.1°C.

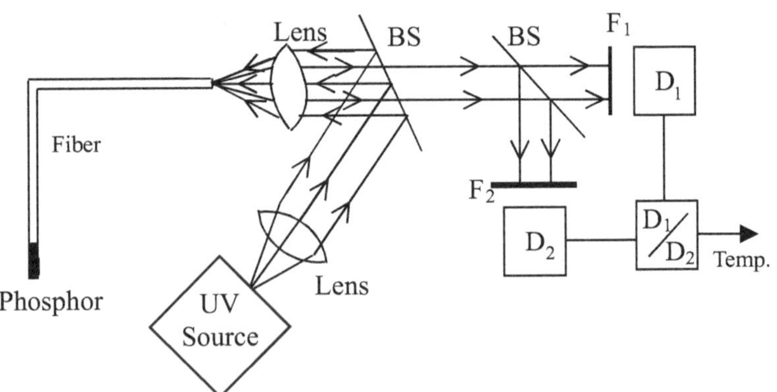

Figure 7.7: (a) Experimental set up of the temperature sensor utilizing temperature dependence of the emission spectrum of phosphor[3].

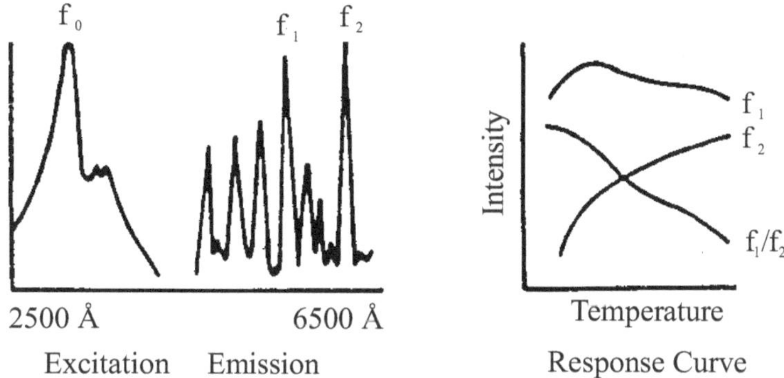

Figure 7.7: (b) Emission spectrum of phosphor and the response curve of the temperature sensor[3].

7.4.4 Semiconductor photoluminescence

Here we describe a temperature sensor based on the temperature dependent photoluminescence of GaAs semiconductor. The sensing element is a double-hetro structure GaAs epitaxial layer surrounded by two $Al_xGa_{1-x}As$ layers. When the GaAs absorbs the incident light, the electron-hole pairs are generated in the GaAs layer. After that, they recombine and re-emit the photons with a wavelength determined by temperature. The luminescent wavelength shifts towards longer wavelengths as the temperature increases. This is due to the decrease in the energy gap, E_g, with temperature. The double hetro structure of the sensing element provides excellent quantum efficiency for the luminescence because the generated electron-hole pairs are confined between the two potential barriers.

The sensor design is shown in Fig. 7.8. The probe is attached at one of the ends of a silica fiber of 100 μm core diameter. Light from an LED (750 nm) is coupled into the fiber and guided to a GRIN lens mounted to a block of glass. One optical interference filter (IF_1) located between the GRIN lens and the glass block, reflects the excitation light which is then guided by the second fiber to the probe. The returned luminescent light is transmitted by the filter IF_1 which is transparent for this wavelength. The reflectivity of the second interference filter (IF_2) changes at about 900 nm. Because the peak wavelength of luminescent light shifts toward longer wavelength with temperature, the ratio between the transmitted and the reflected light intensities of

IF_2 changes. However, the ratio is independent of any variation in the excitation light intensity and any kind of loss. The two lights separated by IF_2 are detected by photodiodes 1 and 2. The temperature range of the sensor is from 0° C to 200° C with an accuracy of ±1° C.

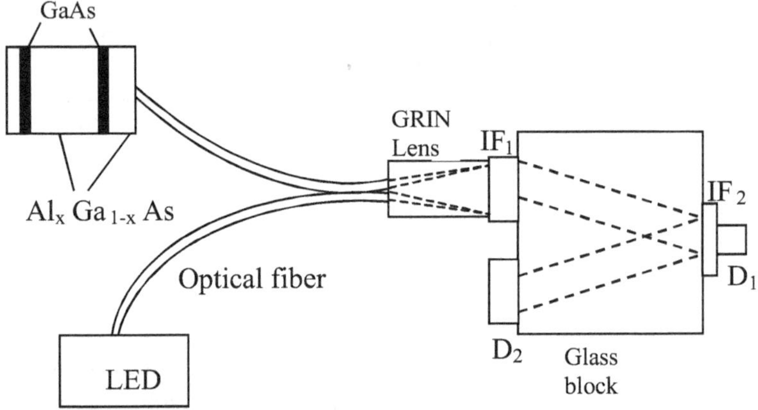

Figure 7.8: Temperature sensor based on temperature dependent photoluminescence from GaAs epitaxial film.

7.4.5 Fluorescence Lifetime

The temperature sensor described above is based on the temperature dependent fluorescence intensity. Here we describe a temperature sensor based on fluorescence decay time[4]. The experimental set up is shown in Fig. 7.9. A small sample of the synthetic laser crystalline material alexandrite, absorbing strongly in the green part of the spectrum, is optically excited by light from an ultra bright, broadband green LED *via* a PCS optical fiber. A short pass optical filter (λ<620 nm) is placed between LED and the fiber to prevent light of longer wavelengths mixing with the fluorescent light produced by the crystal (λ>675 nm). The fluorescent light is collected by a second fiber and transmitted to a photodetection system. A long pass filter is placed in front of the detector to remove any scattered excited light from LED. The crystal is glued to the ends of both the fibers. A thin sheet of aluminium foil is also bonded around the crystal to increase the amount of fluorescence collected by the second fiber.

The LED is modulated with a sinusoidal signal at a frequency of 1KHz which produces sinusoidal light intensity output. This produces a sinusoidal fluorescent response at an identical frequency but shifted in phase. The input modulating signal is used as a reference to establish an arbitrary zero phase position. The size of the phase shift depends on the fluorescent decay time of alexandrite. This decreases with the increase in temperature from room temperature to 150° C. The accuracy of the sensor is less than ± 2° C and its response time is approximately 1 sec.

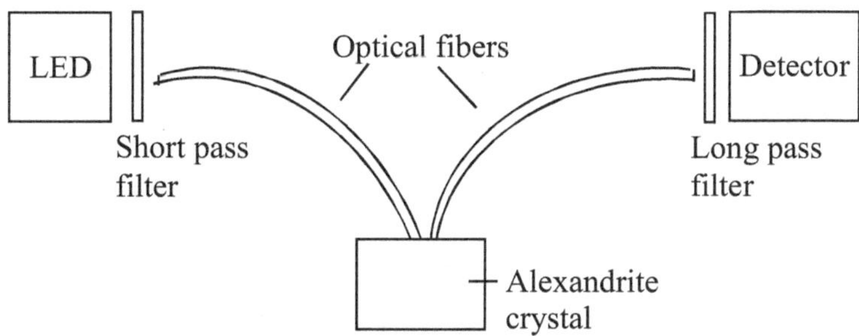

Figure 7.9: Fiber optic temperature sensor based on fluorescence lifetime[4].

There are number of factors which limit the measurement range of fluorescence based temperature sensors. These are stability and the melting point of the fluorescent material, compatibility of the fibers with temperature, the source power and signal processing scheme. The alexandrite material has 1800°C as the melting point. By using silica glass fiber, high power laser diode and a new signal processing scheme the measurement range of the sensor is increased to 700°C with a 1°C accuracy[5].

7.5 Humidity sensor

In (See 6.4; Page 92) we have described humidity sensors based on change in absorption coefficient of the dye by the moisture present around it. Here we shall describe humidity sensor that is based on the change in fluorescence intensity due to the presence of moisture[6]. To fabricate the probe, cladding is removed from a part of the 1 mm diameter glass core UV fiber. The polymethylmethacrylate (PMMA), a polymer, is mixed with a 5×10^{-3} mol/l 1,4-dioxane solution of the umbelliferon (UM) dye at a concentration of 10% (by weight) and is deposited on the glass core by means of a spinner. The thickness of the film prepared is about 1 μm and its pH is around 5.0. Figure 7.10 shows the humidity probe. The UV light (λ<380 nm) is focused onto one of the ends of the fiber. The beam excites the UM dye entrapped within the PMMA clad part of the fiber.

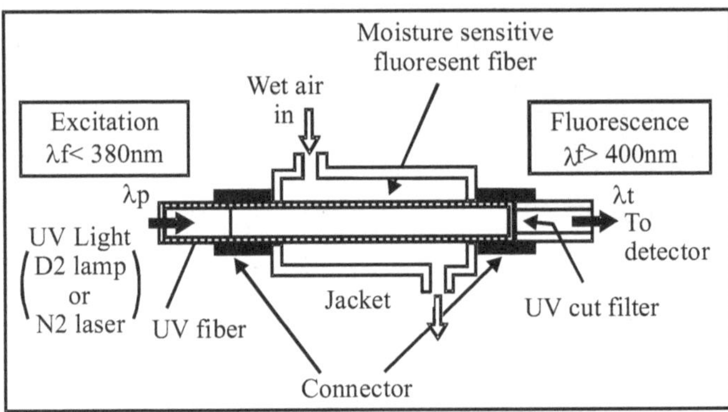

Figure 7.10: Fiber optic humidity probe coated with fluorescent material[6]. Reprinted with permission.

The humidity dependent fluorescence intensity (λ>400 nm) is detected at the other end of the fiber using a monochromator and a photomultiplier tube. The fluorescence intensity increases linearly with relative humidity in the range 20 to 80% RH (see Fig. 7.11). The response of the sensor is very fast. In addition, the sensitivity is not influenced by the presence of toxic gases such as NH_3 and HCl.

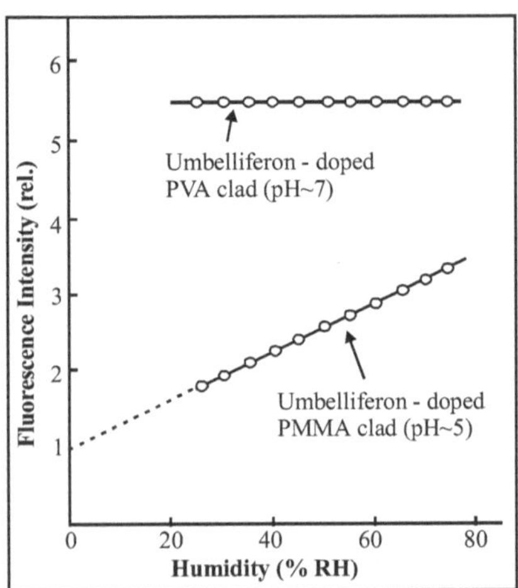

Figure 7.11: Relative fluorescence intensity as a function of atmospheric humidity[6]. Reprinted with permission.

7.6 Glucose sensors

Glucose sensors are required for the determination of glucose in clinical analysis of blood for diagnosis of metabolic disorders (*e.g.* diabetes) and biotechnology. One possible glucose sensor[7] proposed is shown in Fig. 7.12. It is based on the principle of competitive binding. Here first we shall describe competitive binding principle and then describe the glucose sensor based on it.

In a competitive binding based sensor an analyte (A) and a ligand (L) compete for binding to an immobilized reagent (R). The essential feature of ligand is that its optical properties change upon binding to reagent. The reagent and the ligand are confined by a dialysis membrane that allows analyte A to transfer freely between phases. The reactions and associated equilibria can be written as

$$A + R \Leftrightarrow AR\,, \qquad K_A = \frac{[AR]}{[R][A]} \tag{7.9}$$

$$L + R \Leftrightarrow LR \qquad K_L = \frac{[LR]}{[R][L]} \tag{7.10}$$

The total concentrations of L and R in the reagent phase, $[L_o]$ and $[R_o]$ respectively, can be written as

$$\begin{aligned} [L_o] &= [L] + [LR] \\ [R_o] &= [LR] + [AR] + [R] \end{aligned} \tag{7.11}$$

Equations (7.9) and (7.10) give

$$[A] = \frac{[AR][L]}{[LR]}\frac{K_L}{K_A} \tag{7.12}$$

If the equilibrium constants K_A and K_L are large then [R]<<[LR] and [AR]. Thus Eq.(7.12) can be written as

$$[A] = \frac{\left([R_o] - [L_o] + [L]\right)[L]}{[L_o] - [L]}\frac{K_L}{K_A} \tag{7.13}$$

Thus the concentration of the analyte, [A], is related to the concentration of the free ligand, [L].

In the glucose sensor based on above principle, glucose competes for binding sites on a substrate (concanavalin A) with an indicator dye, fluorescein isothiocyanate and dextran (abbreviated as FITC dextran).

The protein or substrate is immobilized or fixed inside a glucose-permeable hollow fiber sealed at one end and the other end is fastened to the end of an optical fiber (Fig. 7.12). Before sealing the end, the small cavity is filled with FITC dextran. The hollow fiber acts as the container and is impermeable to the large molecules. The excitation light transmitted through the fiber sees only the unbound FITC dextran inside the hollow fiber. The fluorescent light produced passes back along the same fiber to a measuring system. In the absence of glucose, a fraction of FITC dextran binds with the immobilized concanavalin A on the membrane surface. If the hollow fiber is immersed in a glucose solution, glucose penetrates into the hollow fiber through the membrane. There comes the competition for the binding sites. Glucose, depending on its concentration, dissociates the FITC dextran from the inner wall of the fiber to the center of the probe. Thus if the glucose concentration increases, free FITC dextran concentration also increases. Thus the fluorescent intensity collected by the fiber increases. Thus the measurement of fluorescent intensity gives the glucose concentration. The sensor has a linear response in the range 50-400 mg glucose per 100 ml.

The response time of the sensor depends on (a) the diffusion rate of the glucose through the wall of the hollow fiber, (b) the competitive displacement reaction rate for the binding sites and (c) the diffusion rate of FITC dextran within the hollow fiber compartment. The step (b) is very fast as compared to steps (a) and (c). The overall time constant of the sensor ranges from 6 to 12 min. The device is stable for a few days but drift occurs over longer time periods.

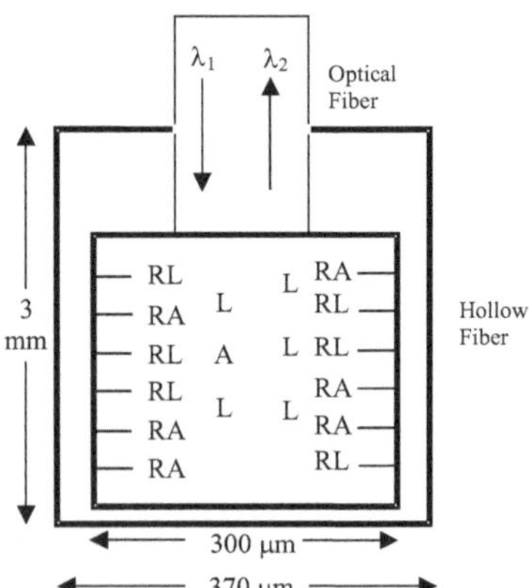

Figure 7.12: Glucose sensor based on competitive binding. R : concanavalin A; A : Glucose; L: FITC.

A modified glucose sensor and based on fluorescence energy transfer was developed by Meadows and Schultz[8]. The FITC-dextran is used as a donor and Rhodamine-labeled Concanavalin A as the acceptor of the energy. As mentioned above the efficiency of the energy transfer process depends on the overlap of the emission spectrum of the donor with the absorption spectrum of the acceptor. In this case there is a relatively good spectral overlap between the two spectra. When FITC-dextran absorbs light of wavelength 470 nm it emits radiation of 520 nm. When FITC-dextran reversibly binds to Rh-con A, it instead of emitting fluorescence intensity transfers its energy to Rh-con A and due to this the fluorescent intensity emission does not occur. The absorption maximum of Rh-con A is at 550 nm. If glucose is added to the system, FITC-dextran is liberated from the Rh-con A. This reduces energy transfer and increases the fluorescence signal.

Figure 7.13 shows the glucose sensor based on energy transfer. The major simplification as compared to the previous sensor is that no immobilization is involved. Light at two different wavelengths, 490 nm and 550 nm are launched alternatively into the fiber and the resulting signals are recorded using a photomultiplier tube. The 550 nm light is launched to know if there is any change in the light source or the detector. This does not affect FITC-dextran. The 490 nm light is used to collect the information regarding the glucose concentration. It is carried out by measuring the fluorescence intensity at 520 nm.

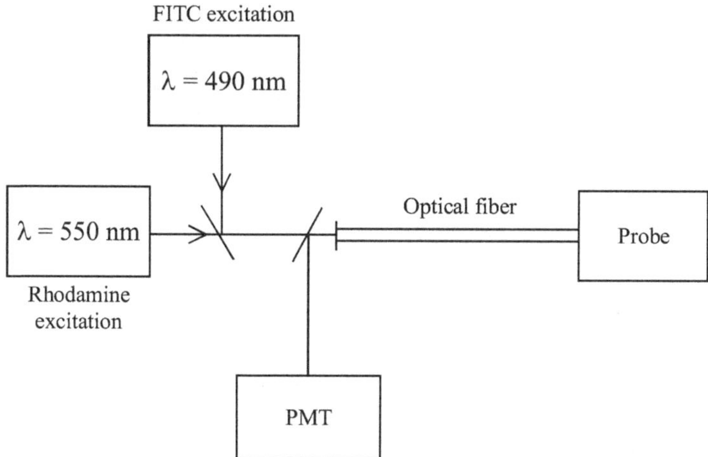

Figure 7.13: Fiber optic glucose sensor based on energy transfer[8].

Another fiber optic glucose sensor[9] consists of an oxygen sensitive fluorescent dye 9,10-diphenylanthracene (9,10-D) and the glucose oxidase (GO_x) enzyme encased in a hydrophilic matrix poly 2-hydroxyethyl-methacrylate (PHEMA). The sensor is fabricated by coaxially grafting PHEMA containing 9,10-D and GO_x to 1 mm diameter fused silica fiber. The thickness of the coating on the fiber can vary from 15 to 600 μm. Glucose and oxygen can both diffuse into the PHEMA coating and undergo the following reaction

$$\text{Glucose} + O_2 \xrightarrow[\text{oxidase}]{\text{glu cos e}} \text{gluconic} - \text{acid} + H_2O_2$$

The glucose probe is shown in Fig. 7.14. Light from an UV source is launched into the fiber. Due to both evanescent wave and scattering, a portion of this light escapes into the PHEMA matrix thus exciting 9,10-D fluorescence. A fraction of this fluorescent light is collected by the same fiber and is transmitted to the detection system. The oxygen inhibits the fluorescence signal. In the presence of glucose, GO_x consumes oxygen in its catalytic cycle thereby lowering the local oxygen concentration which increases fluorescence signal. Thus fluorescent intensity detected is directly proportional to glucose concentration.

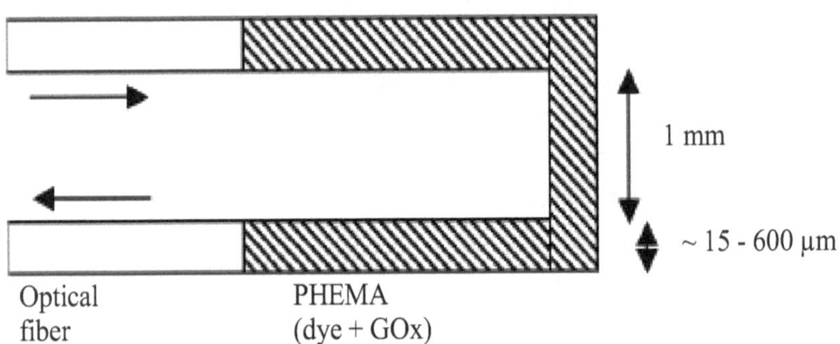

Figure 7.14: Fiber optic glucose probe based on oxygen sensitive fluorescent dye.

The main drawback of this sensor is that at physiological conditions, the glucose concentration may exceed oxygen concentration. In such a situation sensor will not show the actual result. Thus the design of the sensor should be such that a finite level of oxygen remains in the sensor matrix at all possible blood glucose levels.

7.7 pH sensor

In chapter 6 we have described few fiber optic pH sensors based on absorption phenomenon. Here we shall describe fluorescence-based fiber optic pH sensors. The fluorescence-based sensors have advantages as well as disadvantages over absorption-based sensors. The advantages are increased sensitivity, wide dynamic range, wavelength sensitivity and linearity at low concentrations. The disadvantages are the following. First, the excitation radiation or the emission radiation is absorbed by the various other species present in the solution or matrix being analysed. This reduces fluorescence intensity. Second is the identification of an analyte-specific fluorophore. There are a limited number of analyte specific fluorophore.

Figure 7.15 shows the experimental set up of the first fiber optic pH sensor based on fluorescence[10]. It consists of a tungsten halogen lamp, a bifurcated fiber, a photomultiplier tube with housing and a variable slit, a digital photometer and two interference filters. The excitation filter has peak transmittance at 480 nm with a half bandwidth of 7.1 nm while the emission filter's peak transmittance is at 520 nm and its half bandwidth is 8.2 nm. A pH sensitive dye, fluoresceinamine, is immobilized on glass as well as on cellulose. The immobilized fluoresceinamine is then attached to the end of the fiber and is immersed in a beaker containing 15 ml solution of 0.1 M acetic acid. The beaker is covered with a light tight aluminium casing with an injection port. A shutter is placed in front of the detector to exclude ambient light when the aluminium casing is removed. The pH of the solution is varied on adding KOH and is measured by a pH meter. As the pH increases the fluorescence intensity increases up to pH 8.0 for both glass-bound and cellular-bound fluoresceinamine. The response time of the sensor has been reported to be around 15-30 sec. It is slightly high because the film thickness is of the order of 0.1 mm.

Similar to glucose sensor pH sensors also utilize energy transfer mechanism. In energy transfer process, light excites the fluorescent dye which is insensitive to its environment. The fluorescent light is absorbed by an absorber whose absorption spectrum overlaps with the emission spectrum of the fluorescent dye. Thus nonradiative energy transfer occurs from fluorescent material (donor) to an absorber (acceptor). The amount of energy transferred depends upon the environment. Thus the presence of absorber reduces the fluorescent intensity. This phenomenon can also be called a fluorescence-quenching phenomenon. A pH-insensitive eosin, a fluorophore and pH-sensitive phenol red, an absorber are used for the pH sensor[11]. Phenol red is selected because in the pH range 6.0-8.0, it absorbs in the same region

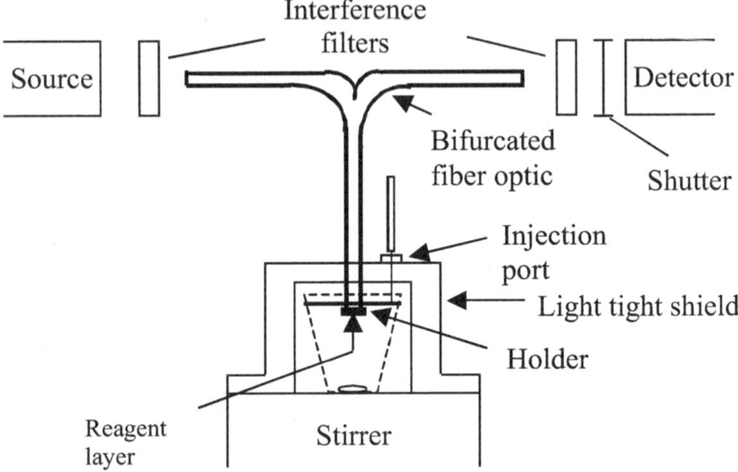

Figure 7.15: Experimental set up of the pH sensor utilizing flouresceinamine dye. Reprinted with permission[10]. Copyright (1982) American Chemical Society.

that eosin fluoresces. Eosin and phenol red are immmobilized on the distal end of an optical fiber. Light from an argon ion laser operating at 488 nm is coupled to the fiber via a neutral density filter and a dichroic mirror. The light excites the eosin and some of the fluorescent intensity emitted from it is absorbed by the phenol red. A part of the remaining fluorescence intensity returns via the same fiber and is deflected 90° by the front surface of the angled dichroic mirror. The fluorescence intensity is detected at 546 nm. As the pH of the fluid around the probe increases the spectral overlap integral between the eosin emission and phenol red absorption spectra increases. This increased overlap results in greater energy transfer from eosin to phenol red. The greater the transfer, the less fluorescence is detected. Thus the fluorescence signal is modulated by the absorber which in turn is modulated by pH. The response time of the sensor is about 0.07 min.

In the fluorescence-based pH sensors described above, a change in fluorescence intensity at a single wavelength is correlated with the presence and amount of analyte (for example H^+). The intensity measurements have a number of drawbacks. These are susceptibility to photobleaching and quenching, variation in dye concentration, inner filter effects, scattering, source fluctuation, susceptibility to temperature variations, and the presence of interfering fluorescent materials and Raman scatter. If the ratio of the fluorescence intensities at two different wavelengths is measured then most of these drawbacks can be removed. The ratio measurement is insensitive to source fluctuations, drift, temperature, quenching and fluophore amount. The inner filter effect and fluorescent interferents effects cannot be eliminated by ratio method. In another method, the presence or amount of analyte is correlated with a change in fluorescence lifetime of a suitable indicator. Lifetime measurement method has several advantages. These include insensitivity to source fluctuations, scattering or absorption of excitation or emission, variation in fluophore levels due to washout or photobleaching, facile calibration, and reduced susceptibility to fluorescent interferents and influences such as temperature. The other advantage of the lifetime approach is the broad dynamic range. Fiber pH sensors based on fluorescence lifetime of indicators have been reported in the literature. One of these sensors[12] has an accuracy of 0.02 pH unit.

All the pH sensors described above are extrinsic type. The first intrinsic type fluorescence based fiber optic pH sensor developed[13] uses evanescent wave excitation of fluorescent dye entrapped in a porous glass cladding prepared by sol-gel method described in (Sec.6.3; Page 89). The sol-gel method allows the preparation, at low temperatures, of a highly transparent glass which can act as a support for chemically sensitive dyes. The tetraethylorthosilicate is used as the precursor and the pH-sensitive dye used is fluorescein. The experimental set up of the pH sensor is shown in Fig. 7.16. A collimated beam of light from an Argon ion laser operating at 488 nm is reflected by a dichroic filter and then coupled into the coated fiber which is vertically mounted in a sample chamber. The evanescent field of the guided radiation excites the entrapped dye. A fraction of the resultant fluorescence is coupled back into the fiber and some of this signal propagates back through the dichroic filter to the monochromator and detector. Figure 7.17 shows the measured fluorescence intensity at $\lambda = 530$ nm as a function of pH. It can be seen that the greatest sensitivity to pH change occurs in the region pH 3.5 to 6.5. The response time of the sensor is reported to be approximately 5 sec.

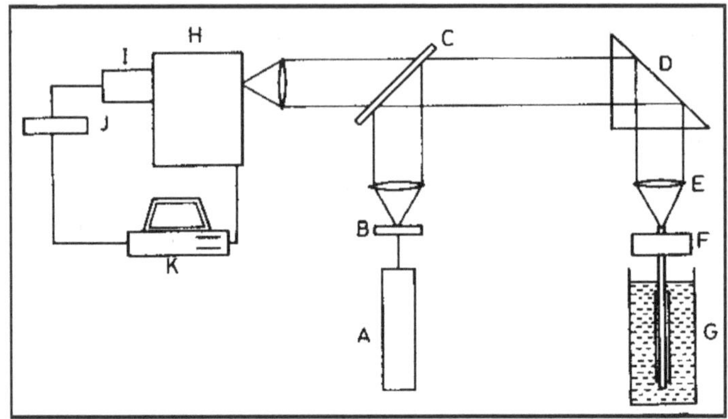

Figure 7.16: Intrinsic type fiber optic pH sensor[13]. Reprinted with permission. À© [1991] IEEE.

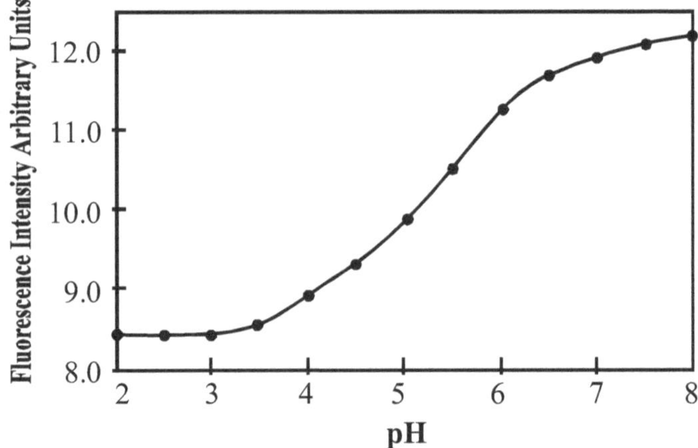

Figure 7.17: Calibration curve of the pH probe prepared using sol gel method[13]. Reprinted with permission. À© [1991] IEEE.

7.8 Oxygen sensor

In medical field, the oxygen partial pressure (PO_2) of blood and tissue are measured by either in vivo optical measurement of hemoglobin saturation or by a transcutaneous sensor or by withdrawing sample for a blood gas measurement instrument. The disadvantages of these methods are that these are not practical, reliable and safe. Fiber optic sensors have also been developed for oxygen determination. These sensors have fast response, do not consume oxygen, and are not easily poisoned. Most of the fiber optic oxygen sensors are based on the quenching of fluorescence emitted by some chemical species in the presence of oxygen. One such sensor for in vivo measurement of oxygen partial pressure of blood and based on fluorescence quenching[14] is shown in Fig.7.18. The sensor is small in size. It can be inserted into a blood vessel or tissue in various parts of the body with little disturbance. It can also be used for continuous monitoring of PO_2 levels. The general construction is similar to the fiber optic pH sensor shown in Fig.6.9. To fabricate the sensor plastic optical fibers are used. This is because these are highly flexible and do not present the risk of breakage in sharp bends. The probe at the end of fibers consists of a dye, perylene dibutyrate, adsorbed on macroreticular polystyrene adsorbent contained in a porous polypropylene envelope. The dye has an excitation peak at 468 nm and emission peak at 514 nm. The dye is nontoxic and has a sufficient oxygen quenching sensitivity. The fibers are of 250 μm diameters while the probe is 3 mm long and 0.6 mm in diameter. Blue light is transmitted through one of the two fibers. The light exited from this fiber illuminates the dye that emits a green fluorescent light upon excitation. The intensity of the fluorescent light depends on the partial pressure of oxygen. The green fluorescent light and the scattered blue light are returned through the other fiber for transmission to the detecting system. The intensity of the fluorescent light decreases with the increase in the level of oxygen that passes through the gas-permeable membrane and reacts with the dye. The following relation (Stern-Volmer relation) is used to calculate the value of PO_2

$$\frac{I_0}{I} = 1 + \frac{PO_2}{P} \tag{7.14}$$

where I_0 is the fluorescence intensity without oxygen quenching (i.e. in the absence of oxygen or $PO_2 = 0$), I is the fluorescence intensity when the oxygen partial pressure is PO_2 and P is the partial pressure at half quench and is a constant. The ratio of the intensities (I_0/I) is measured electronically to obtain the value of PO_2. The performance of the probe is tested in a gas stream, in water in a temperature regulated cell and in vivo. The sensor can be used for the range 0-150 torr with a precision of 1 torr (1torr = 1 mm Hg). The stability of the sensor over a long time is good but does not allow indefinite storage. After few years a similar kind of fiber optic PO_2 sensor based on fluorescence quenching was reported[15]. In that sensor pyrenebutyric acid was used as a dye. The dye has excitation peak at 342 nm while emission peak is at 395 nm.

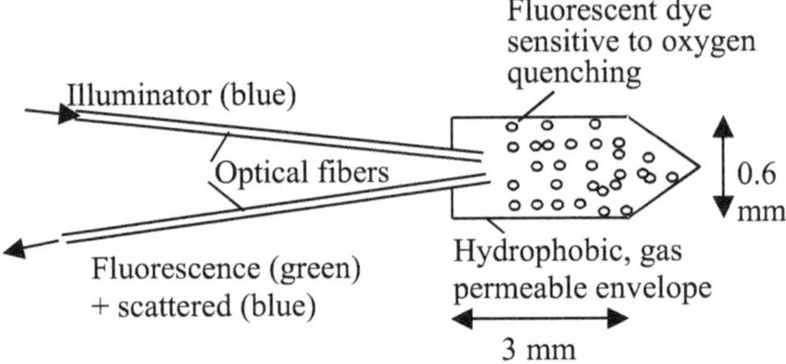

Figure 7.18: Fiber optic PO_2 sensor[14].

Later an intrinsic type oxygen sensor utilizing evanescent wave excitation of the dye immobilized on the core of the fiber using sol-gel process was reported[16]. The probe and the experimental set up of the sensor are shown in (Fig. 7.19). A ruthenium complex, Ru(II) tris

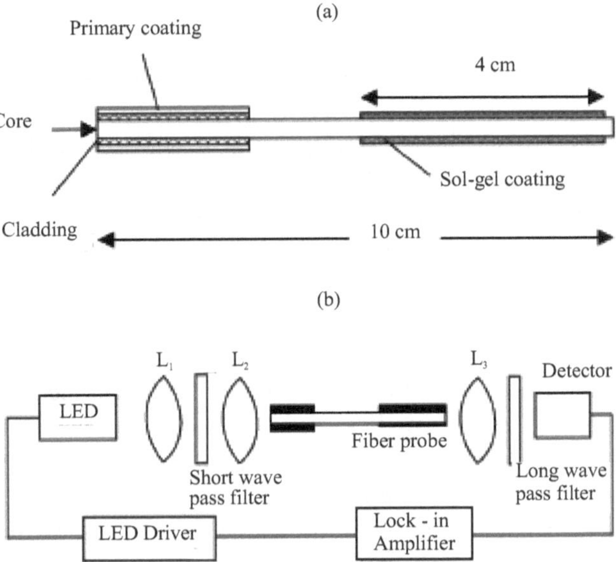

Figure 7.19: Fiber optic oxygen sensor based on evanescent wave excitation. (a) Probe geometry and (b) experimental set up. Adapted from ref.16. À© [1994] IEEE.

(4,7-diphenyl-1,10-phenanthroline) abbreviated as Ru $(Ph_2Phen)_3^{2+}$, is chosen as the indicator for oxygen sensing. The indicator has high sensitivity and strong absorption in the blue-green region of the spectrum for which a range of suitable laser and LED sources are available. The dye is coated on 4 cm length of the unclad fiber using sol-gel process. Light from a blue LED whose spectral output peaks at 450 nm and has considerable overlap with the absorption spectrum of the ruthenium complex is launched into the fiber after passing through a short-wave-pass filter (λ_{cutoff} = 505 nm). The fiber is mounted in a gas cell. The evanescent field of the light propagating excites the entrapped ruthenium complex and a fraction of the resultant fluorescence is captured by the fiber. The forward propagating portion of this fluorescence passes through a long-wave pass filter (λ_{cutoff} = 550 nm) to a silicon photodiode detector. The output from the detector is passed to the lock-in amplifier for synchronous detection with the LED pulsing signal. For the measurements, precise mixtures of nitrogen and oxygen are passed *via* mass-flow controllers to the gas cell. The measurements are carried out for a range of oxygen concentration. The sensor response time is less than 5 sec.

7.9 Carbon dioxide sensor

Figure 7.20 shows the probe of a fluorescence based fiber optic CO_2 sensor[17]. It consists of 14 fibers, 4 in the center and 10 in a symmetrical circle surrounding the central bundle. The 4-fiber bundle in the center carries the input light to the end of the probe while the surrounding 10 fibers guide the modulated light to the PMT for detection.

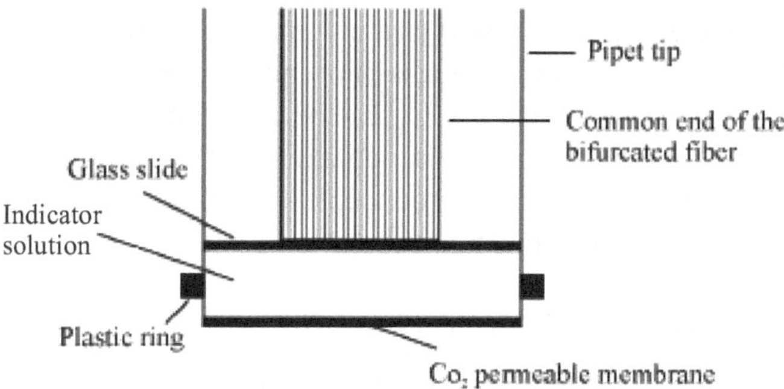

Figure 7.20: A fiber optic CO_2 probe.

The fibers are tightly packed with black shrink tubing at both the single ends and the common end and then are polished to a flat surface. The common end is fitted into the cross section of a pipet tip. A round glass plate is tightly fitted at the end of the fiber bundle. The indicator solution is trapped between the glass plate and the CO_2-gas permeable membrane which is held tightly over the end of the tip by a plastic ring. The indicator solution is prepared as an aqueous solution containing sodium bicarbonate, 1-hydroxypyrene-3,6,8-trisulfonate (HPTS) and sodium chloride. HPTS is a fluorescent indicator having high quantum yield, high absorbance and excellent photostability. Sodium chloride is used to balance the osmolarity of the indicator solution and that of the buffer solution across the membrane. Light from a broadband source is directed onto a heat-absorbing filter, a 460 nm interference filter and the chopper before coupling into the four fiber bundle. Chopper is used to reduce the noise from ambient irradiation. The light exiting from these fibers excites the indicator. Part of the fluorescent light emitted is collected by 10 fibers bundle. It is then passed through a 520 nm interference filter and then to the PMT. The current generated in the PMT is detected by a lock-in amplifier. The measurements are carried out with buffer solutions containing different concentrations of dissolved CO_2. When the probe is dipped into the buffer solution, the dissolved CO_2 diffuses through the membrane until an equilibrium is reached between the partial pressure of CO_2 in the sample solution and the CO_2 in the indicator solution. The CO_2 that diffuses across the membrane decreases the pH of the sensing layer and inhibits the dissociation of the fluorescent indicator. The ratio of fluorescence intensities (I_0/I) in the absence (I_0) and presence (I) of CO_2 is determined for different CO_2 concentration. As the concentration of CO_2 increases the ratio increases. By choosing right amount of indicator, appropriate bicarbonate concentration and a thin enough sensing layer, a linear relationship between I_0/I and $[CO_2]$ can be achieved. The sensor response time varies from 1.5 to 3.5 min depending on the CO_2 concentration measured and the bicarbonate concentration in the indicator solution. The lowest CO_2 concentration measured is 0.2 mM.

7.10 Aluminium (III) sensor

The aluminium sensor described below[18] is similar to the pH sensor shown in Fig.7.15. To sense Al^{3+}, morin is immobilized on a cellular support. Morin is non fluorescent by itself. When Al^{3+} comes in its contact morin-aluminium complex forms. The complex is highly fluorescent. The fluorescent signal depends on the amount of aluminium bound to the morin. The immobilized morin is attached to the end of the bifurcated optical fiber. Light of 420 nm wavelength is used to excite the morin-aluminium complex and the fluorescent light emitted is detected at 488 nm. The aluminium concentration is varied in acetate buffer solution. To measure the fluorescence signal at each concentration the probe is immersed in the solution. The fluorescence intensity increases with the increase in the aluminium concentration. The response of the sensor is linear in the beginning but saturates at higher concentration of Al^{3+}. The saturation occurs due to the nonavailability of free morin molecules. The response curve of the sensor depends on pH. The optimum pH for the sensor is 4.8. The response time of the sensor is 1-2 min. The detection limit is 1×10^{-6} M.

7.11 Immunodiagnostic sensor

There are more than 100 analytes that can be measured by immunodiagnostics. The number will continue to grow with the need to detect and measure concentrations of new pharmaceuticals and the growing number of environmental pollutants and toxins. Laboratory tests are carried out by hospitals to determine the concentration of particular chemicals and biochemicals in blood samples in order to diagnose disease states. The test is based on the specific binding of large molecular weight proteins called antibodies (Ab) with the chemicals to be detected called antigens (Ag) in the sample. In general, there is a reversible reaction of the form

$$Ab + Ag \Leftrightarrow AbAg$$

Due to extremely high equilibrium association constants attainable with antibodies (10^{10} M^{-1} or greater), and with the advent of monoclonal antibody technology providing a wide range of reagent grade immunochemicals antibody-based systems permit detection down to 10^{-12} M of antigens (*i.e.* analytes). The conventional methods used for detection are slow and cumbersome and require large volumes of reagents for the analysis. Further these methods cannot be used for remote sensing. Optical fibers have received considerable attention for the development of immunosensors based on fluorescence phenomenon.

Both intrinsic and extrinsic types of sensors have been developed. We first discuss an intrinsic type sensor utilizing evanescent wave approach.

There are number of methods which can be used to measure, intrinsically, the concentration of an antigen (Ag) in a complex solution containing many other chemicals. As mentioned, antibodies are proteins and most of the proteins are intrinsically fluorescent in UV light. To

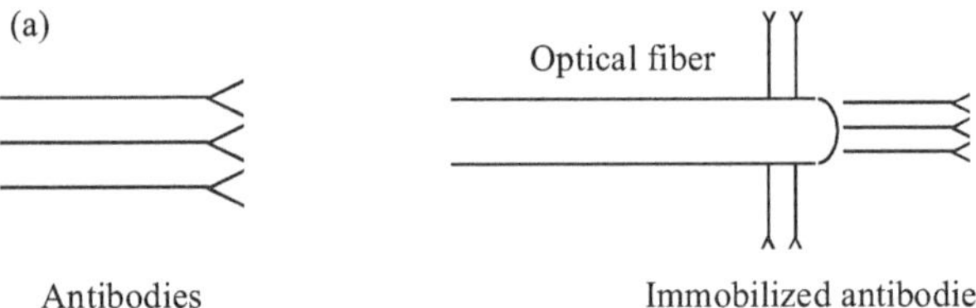

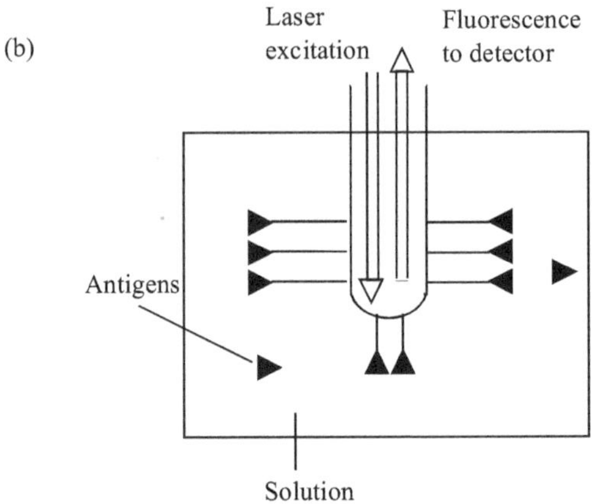

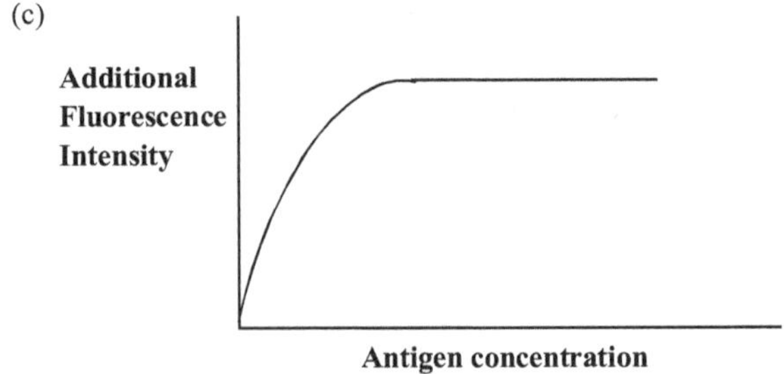

Figure 7.21: (a) Antibodies are immobilized on the unclad fiber tip, (b) immersing immobilized part of the fiber into solution binds fluorescent antigens to antibodies and (c) response characteristics of the sensor.

detect antigen an antibody should be available which can form a complex with it. Suppose that the corresponding antibody is available. These antibodies are immobilized on the surface of an unclad fiber (Fig. 7.21(a)). If UV light is sent through the fiber, the evanescent wave will excite antibodies which produce fluorescence signal. The fluorescence signal is transmitted back through the same fiber to the detector. If the antigen to be detected is also fluorescent then if the fiber tip is inserted into the complex solution (Fig.7.21(b)), the specific antigen present will bind to the immobilized antibody. The evanescent wave will thus excite the antigen and additional fluorescence will be observed. The intensity of the additional fluorescence is proportional to the concentration of antigen in solution as long as immobilized concentration of the antibody is more than the concentration of antigen (see Fig.7.21(c)). If the concentration of antigen is more than that of antibody the signal saturates. This is the direct method to sense antigen. If antigen is not intrinsically fluorescent then a second antibody can be added to sense antigen. The second antibody binds to the antigen which is already bound to the immobilized antibody (Fig.7.22(a)). This increases the fluorescent signal which is related to the amount of bound antigen. The response curve is shown in Fig.7.22(b). If the second antibody is labeled with an extrinsic fluor, such as fluorescein or rhodamine, one can still detect antigen (Fig.7.23(a)). Here also fluorescent intensity increases with the concentration of antigen (Fig. 7.23(b)). The advantage is that by using proper fluor, blue-green excitation light can be used instead of UV light. This simplifies optics and light source requirement.

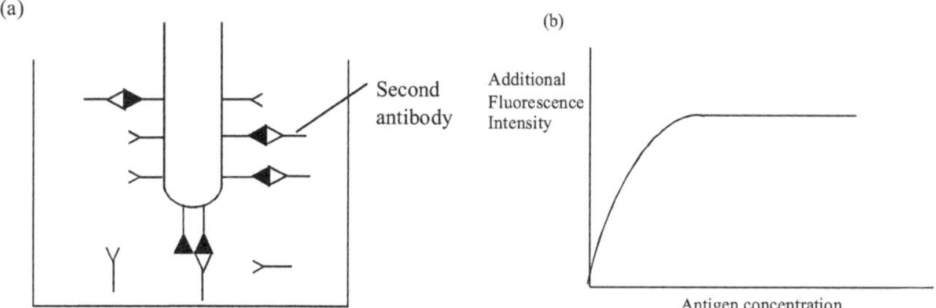

Figure 7.22: (a) Immersion of fiber into solution containing non-fluorescent antigens. Additional antibodies are added. (b) response characteristics of the sensor

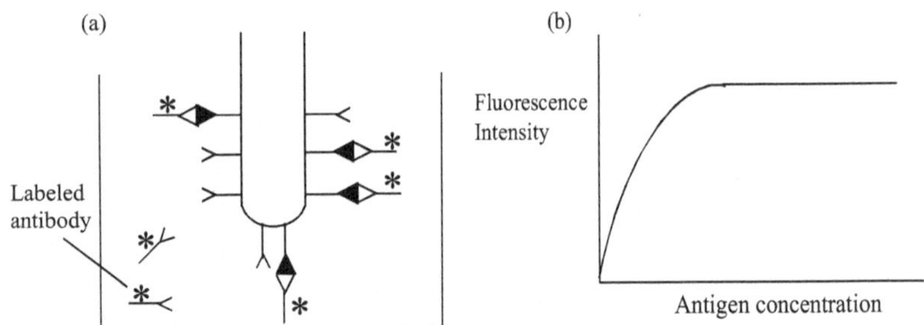

Figure 7.23: (a) Immersion of fiber into the solution containing labeled antibody. (b) Response characteristics of the sensor.

The antigen concentration can also be determined using competitive binding method. In this case fluorescently labeled antigens are added in the solution. The labeled antigen competes with unlabeled antigen in the solution for the surface-bound antibody Fig.7.24(a). The more the bulk antigen in the solution, the less labeled antigen can bind and *vice versa*. Therefore, the detected fluorescence intensity decreases with the increase in the concentration of the antigen Fig.7.24(b). Such a competitive binding method has been extensively used in many chemical sensors and is very sensitive and specific

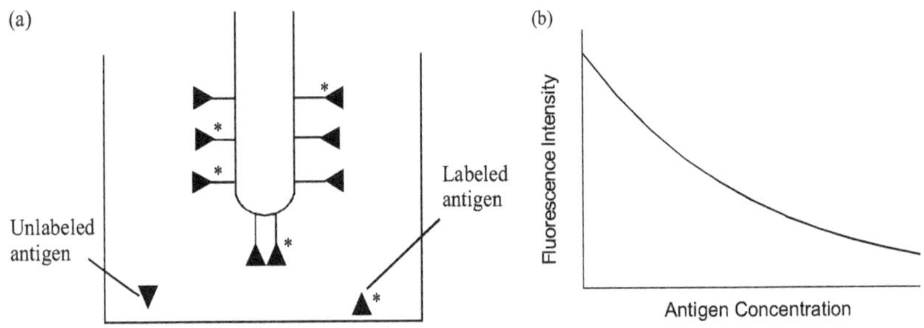

Figure 7.24: (a) Immersion of fiber into the solution containing labeled antigen. (b) Response characteristics of the sensor.

The experimental set up to detect the fluorescent signal is shown in Fig.7.25. Collimated beam from a light source is launched into the fiber via beam splitter and objective. The fluorescent light is collected by the fiber and is detected by the detector after passing through the objective, beam splitter, lens and the corresponding interference filter. The sensor can be found useful in detecting viral infections such as the AIDS virus, screening for cancer, assessing human exposure to toxic chemicals, monitoring response to drug therapy *etc.* The difficulties with this futuristic technique are selection of appropriate antibodies, immobilizing them on the fiber and their reversibility. Further antibody-antigen binding can be disrupted by changes in pH, ionic strength, temperature or local water structure.

Figure 7.26 shows an extrinsic immunosensor. This is based on the competitive binding reaction. The antibody is covalently bound to the inner wall of a hollow fiber. The fluorescently labeled antigen is kept inside the hollow fiber. The porosity of the hollow fiber is that unlabeled antigen can pass through freely while the labeled antigen is retained inside. The hollow fiber is attached at the end of an optical fiber with low numerical aperture. Initially labeled antigens are bound to the antibody. As the concentration of unlabeled antigen increases outside the hollow fiber it penetrates into the hollow fiber and labeled antigen is displaced from antibody binding sites.

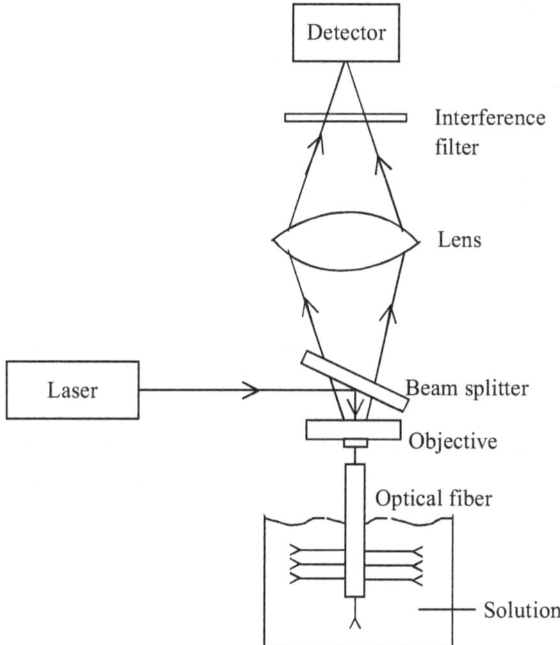

Figure 7.25: Fiber optic immunosensor configuration.

The excitation light launched in the fiber sees labeled antigen and excites them. The fluorescent signal generated is coupled to the same fiber and is detected. The fluorescent signal increases as the concentration of unlabeled antigen increases.

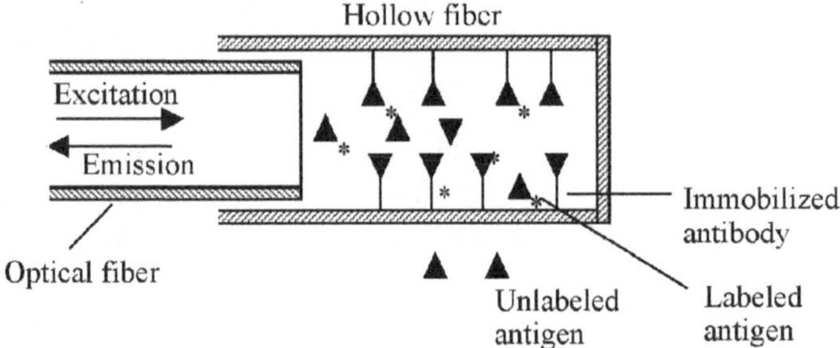

Figure 7.26: Extrinsic type fiber optic immunosensor with immobilized antibody.

In a second type of extrinsic sensor antigen is immobilized to the inner wall of the hollow fiber (Fig.7.27). The fluorescently labeled antibody is kept inside the hollow fiber. Initially labeled antibodies are bound to immobilized antigen.

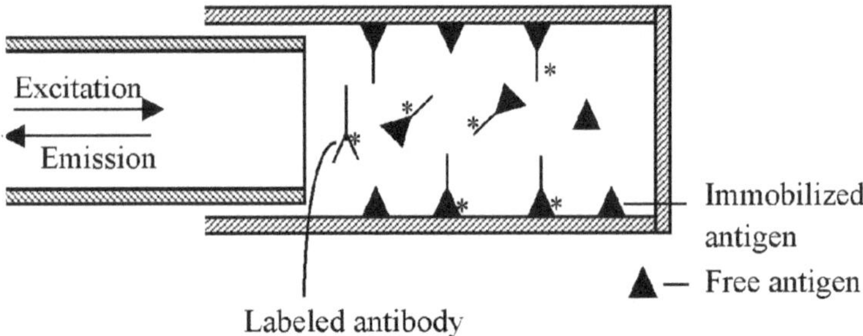

Figure 7.27: Extrinsic type fiber optic immunosensor with immobilized antigen.

When the probe comes in contact of antigen, to be detected, they penetrate inside the hollow fiber and binds with the antibody. As the concentration of antigen increases more and more antibodies bind to the non-immobilized antigen. This increases the fluorescence intensity.

References

1. K. Kyuma, S. Tai, T. Sawada and M. Nunoshita (1982) Fiber optic instrument for temperature measurement. *IEEE J. Quant. Electron. QE* **18**, 676-679.
2. T. Nakayama (1984) Optical sensing techniques by multimode fibers. *SPIE Proc*. **478**, 19-26.

3. K.A. Wickersheim (1987) A new fiber optic thermometry system for use in medical hypothermia. *SPIE Proc.* **713**, 150-157.

4. A.T. Augousti, K.T.V. Grattan and A.W. Palmer (1988) Visible-LED pumped fiber optic temperature sensor. *IEEE Trans. Instrument. Measurement* **37**, 470-472.

5. Z. Zhang, K.T.V. Grattan and A.W. Palmer (1992) Fiber-optic high-temperature sensor based on the fluorescence lifetime of alexandrite. *Rev. Sci. Instrum.* **63**, 3869-3873.

6. S. Muto, A. Fukasawa, M. Kamimura, F. Shinmura and H. Ito (1989) Fiber humdity sensor using fluorescent dye-doped plastics. *Jap. J. Appl. Phys.* **28**, L1065-L1066.

7. J.S. Schultz, S. Mansouri and I.J. Goldstein (1982) Affinity sensor: a new technique for developing implantable sensors for glucose and other metabolites. *Diabetes Care* **5**, 245-253.

8. D. Meadows and J.S. Schultz (1988) Fiber-optic biosensors based on fluorescence energy transfer. *Talanta* **35**, 145-150.

9. R. Shah, M. Gold and R.M. Schaffer (1988) Development of a fiber optic oxygen and glucose sensor. *IEEE Engineering in Medicine & Biology Society 10th Annual Conference*, pp 742-744.

10. L.A. Saari and W.R. Seitz (1982) pH sensor based on immobilized fluoresceinamine. *Anal. Chem.* **54**, 821-823.

11. D.M. Jordan, D.R. Walt and F.P. Milanovich (1987) Physiological pH fiber-optic chemical sensor based on energy transfer. *Anal. Chem.* **59**, 437-439.

12. R.B. Thompson and J.R. Lakowicz (1993) Fibre optic pH sensor based on phase fluorescence lifetimes. *Anal. Chem.* **65**, 853-856.

13. B.D. MacCraith, V. Ruddy, C. Potter, B. O'Kelly and J.F. McGilp (1991) Optical waveguide sensor using evanescent wave excitation of fluorescent dye in sol-gel glass. *Electron. Lett.* **27**, 1247-1248.

14. J.I. Peterson, R.V. Fitzgerald and D.K. Buckhold (1984) Fiber-optic probe for in vivo measurement of oxygen partial pressure. *Anal. Chem.* **56**, 62-67.

15. D.W. Lubbers and N. Optiz (1983) Optical fluorescence sensors for continuous measurement of chemical concentrations in biological systems. *Sensors and Actuat.* **4**, 641-654.

16. B.D. MacCraith, G. O'Keeffe, C. McDonagh and A.K. McEvoy (1994) LED-based fiber optic oxygen sensor using sol-gel coating. *Electron. Lett.* **30**, 888-889.

17. X. He and G.A. Rechnitz (1995) Linear response function for fluorescence-based fiber-optic CO_2 sensors. *Anal. Chem.* **67**, 2264-2268.

18. L.A. Saari and W.R. Seitz (1983) Immobilized morin as fluorescence sensor for determination of aluminium (III). *Anal. Chem.* **55**, 667-670.

8

Interferometric Sensors

8.1 Introduction

Interferometric or phase modulated sensors have received special attention because these are much more sensitive than intensity modulated sensors and can be used over a much larger dynamic range. The sensors rely on the interferometric technique. Generally, the sensor employs a coherent light source such as a laser and two single mode optical fibers. Light is split and coupled into both the fibers. If the measurand perturbs the optical properties of one of the fibers relative to the other, a phase shift occurs. The change in phase can be measured very precisely. For the measurement a sophisticated electronic processing equipment is needed. Thus the interferometric sensors are expensive compare to intensity modulated sensors. There are four kinds of interferometer configurations: Mach-Zehnder, Michelson, Fabry-Perot and Sagnac interferometers. The Mach-Zehnder, Michelson and Sagnac interferometers are the two beam interferometers while in the case of Fabry-Perot interferometer multiple beam interference occurs. In this chapter we shall first describe interference phenomenon and then different kinds of fiber optic interferometers. In the end we shall describe a number of sensors based on interference phenomenon or phase modulation.

8.2 Interference phenomenon

When two or more waves superpose one obtains an intensity distribution which is known as the interference pattern. A stationary interference pattern is observed when the interfering waves maintain a constant phase difference. One does not observe a stationary interference pattern between the waves emanating from independent sources due to the very process of emission which does not maintain the constant phase difference between the waves. Thus, to observe interference pattern, it is necessary that the interfering waves originate from a single source. Below we shall describe the production of two or more waves from a single source and their interference patterns.

8.2.1 Two beam interference

There are two methods for obtaining two coherent beams. One is the division of wavefront and the other is the division of amplitude of the incident beam. The division of wavefront method for obtaining two coherent beams is shown in Fig. 8.1. The source S emits radiation in all directions but only two separated parts of the beam are used. These two beams are then superimposed in the region in which the interference phenomenon is to be observed. Its examples are Young's double slits, Fresnel's mirrors, Lloyd's mirror *etc.* In the case of division of amplitude, the incident beam falls on a semitransparent plate as shown in Fig. 8.2. One part of the beam is transmitted and the other part is reflected. The two beams are then superposed in the region in which the interference phenomenon is observed. The examples are Michelson interferometer, Newton's ring *etc.*

Consider two monochromatic waves represented by E_1 and E_2 and propagating in the same direction. Mathematically, the amplitudes of these waves can be written as

$$E_1 = a_1 \exp[i(\omega t - \phi_1)] \tag{8.1}$$

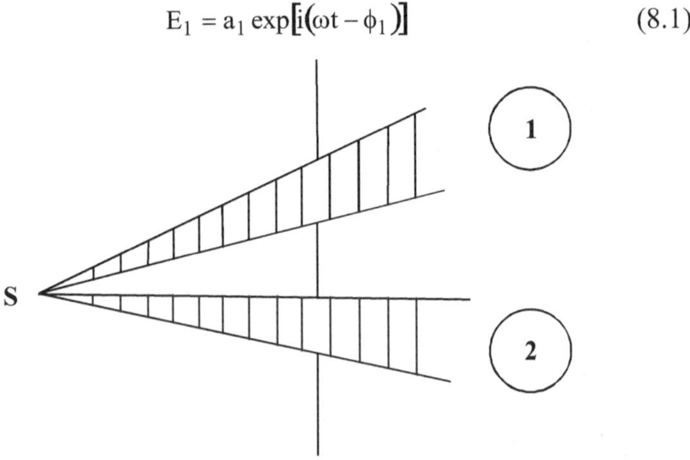

Figure 8.1: Production of two coherent beams by division of wavefront.

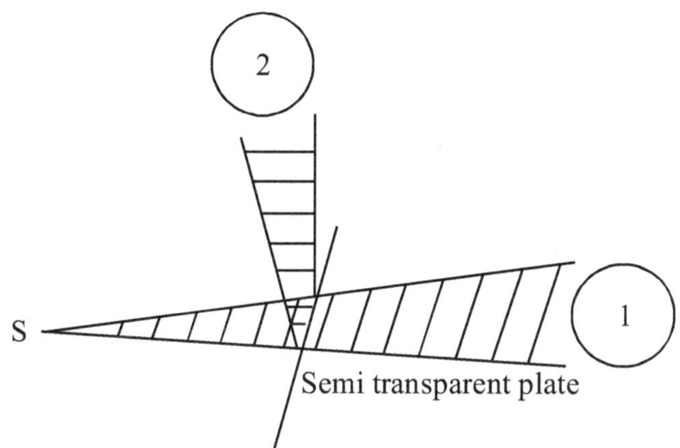

Figure 8.2: Production of two coherent beams by division of amplitude.

and

$$E_2 = a_2 \exp[i(\omega t - \phi_2)] \tag{8.2}$$

where ω is the angular frequency and ϕ_1 and ϕ_2 are the phases. The superposition of these two waves at a point gives

$$E = E_1 + E_2 = [a_1 \exp(-i\phi_1) + a_2 \exp(-i\phi_2)]\exp(i\omega t) \tag{8.3}$$

The intensity at that point is given by

$$\begin{aligned} I = EE^* &= a_1^2 + a_2^2 + 2a_1 a_2 \cos\phi \\ &= I_1 + I_2 + 2\sqrt{I_1 I_2}\cos\phi \end{aligned} \tag{8.4}$$

where $\phi = \phi_1 - \phi_2$, is the phase difference between two waves at the point of observation, I_1 and I_2 are the intensities at the point due to two waves acting separately. The constructive interference (maximum intensity) occurs when $\cos\phi = 1$ or $\phi = 2k\pi$, k = 0, 1, 2, 3, —, and the intensity is given by

$$I_{max} = I_1 + I_2 + 2\sqrt{I_1 I_2} \tag{8.5}$$

On the other hand, when $\cos\phi = -1$ or $\phi = (2k+1)\pi$, the destructive interference (minimum intensity) occurs and the corresponding intensity is

$$I_{min} = I_1 + I_2 - 2\sqrt{I_1 I_2} \tag{8.6}$$

For $I_1 = I_2 = I_0$, $I_{max} = 4I_o$ and $I_{min} = 0$. A plot of irradiance I versus phase difference ϕ is shown in Fig. 8.3. It exhibits periodic variation of irradiance and hence fringes. It may be noted that energy is not conserved at each point of the superposition, *i.e.*, $I \neq 2I_o$, but for one spatial period of the fringe pattern, $I_{av} = 2I_o$. The power falls at some points but rises at other points such that the total pattern satisfies the principle of energy conservation. For the case of interfering beams of equal amplitude, the intensity is given by

$$I = 2I_0(1 + \cos\phi) = 4I_0 \cos^2\left(\frac{\phi}{2}\right) \tag{8.7}$$

where

$$\phi = \frac{2\pi}{\lambda}\delta,$$

δ is the path difference between the two beams interfering at the point of observation.

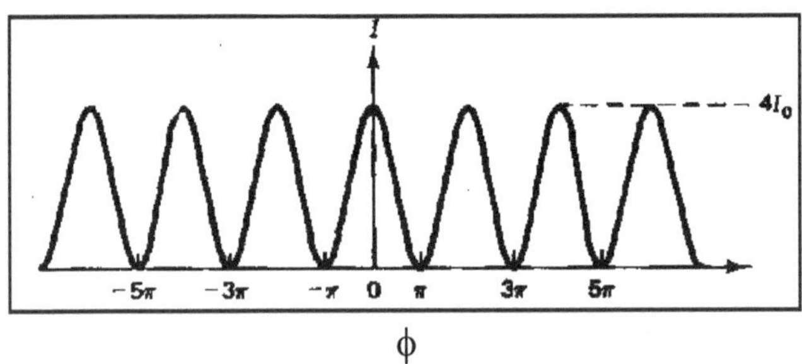

Figure 8.3: Irradiance versus phase difference for two beam interference.

8.2.2 Multiple beam interference

In this section we shall discuss the interference phenomenon involving many beams that are derived from a single beam by multiple reflections and hence using division of amplitude. Let us consider a plane parallel film of thickness e and of refractive index n surrounded by air and illuminated by a plane wave at a finite angle (see Fig. 8.4). The wave will suffer multiple reflections at the two parallel surfaces of the film. Let A_o be the complex amplitude of the wave incident on the film at an angle i. If ϕ represents the phase difference between two successive waves emanating from the film then

$$\phi = \frac{2\pi}{\lambda}\delta = \frac{4\pi n e \cos i'}{\lambda}$$

where i' is the angle of refraction inside the film. Let t and r represent the amplitude transmission and reflection coefficients when the wave is traveling from the surrounding medium into the film and t' and r' represent the corresponding quantities when the wave is traveling from the film into the surrounding medium. These coefficients have the following relationships

$$r = -r' \tag{8.8}$$

$$1 - tt' = r^2 = (r')^2 \tag{8.9}$$

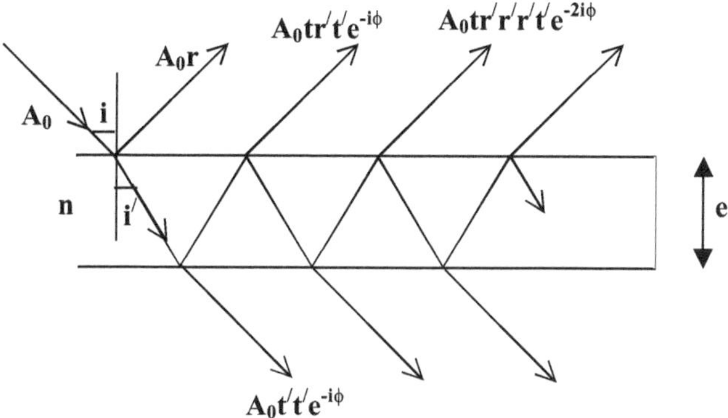

Figure 8.4: Reflection and transmission of a beam incident on a film of uniform thickness.

If R and T represent the energy reflectivity and transmitivity of the film surfaces then

$$R = r^2 \tag{8.10}$$

$$T = tt' \tag{8.11}$$

and hence

$$R + T = 1 \tag{8.12}$$

In the above equation we have assumed that there is no absorption of light in the film. The amplitudes of the successive reflected waves from the film will be

$$A_or,\ A_otr't'e^{-i\phi},\ A_otr'r'r't'e^{-2i\phi},\ ....$$

If the reflectivity is small *i.e.* $|r| = |r'| << 1$, then the amplitudes of the first two waves are approximately equal and the other waves have negligible amplitudes. In that case we shall have two beams interference. If the reflectivity is large then all the reflected waves will have comparable amplitudes. In that case we shall have multiple beam interference. Further, if the film is sufficiently long and the angle of incidence i is small then there will be a large number of waves and we can assume them to be infinite in number. Thus the resultant amplitude of the reflected wave would be

$$A_r = A_o r + A_o t r' t' e^{-i\phi} + A_o t r'^3 t' e^{-2i\phi} + \text{------}$$
$$= A_o[r + t t' r' e^{-i\phi} \{1 + r'^2 e^{-i\phi} + r'^4 e^{-2i\phi} + \text{-------}\}] \quad (8.13)$$

Use of Eqs.(8.10) and (8.11) gives

$$A_r = A_o r[1 - T e^{-i\phi} \{1 + r^2 e^{-i\phi} + r^4 e^{-2i\phi} + \text{------}\}]$$
$$= A_o r\left[1 - \frac{Te^{-i\phi}}{1 - Re^{-i\phi}}\right] \quad (8.14)$$

Using Eq. (8.12) we obtain

$$A_r = A_o r\left[\frac{1 - e^{-i\phi}}{1 - Re^{-i\phi}}\right] \quad (8.15)$$

The intensity of the reflected beam is given by

$$I_r = A_r A_r^* \quad (8.16)$$

Substitution of Eq.(8.15) in Eq.(8.16) gives

$$I_r = A_o^2 R\left[\frac{2(1 - \cos\phi)}{1 + R^2 - 2R\cos\phi}\right]$$

$$= I_o\left[\frac{F\sin^2\frac{\phi}{2}}{1 + F\sin^2\frac{\phi}{2}}\right] \quad (8.17)$$

where I_o $(= A_o^2)$ is the intensity of the incident beam and $F = 4R / (1-R)^2$, is called the coefficient of finesse. The intensity transmitted through the film is given by

$$I_t = I_o - I_r = \frac{I_o}{1 + F\sin^2\frac{\phi}{2}} \quad (8.18)$$

For the transmitted beam constructive interference occurs when $\sin^2(\phi/2) = 0$ or $\phi = 2k\pi$, k = 0, 1, 2, -------. In that case the maximum intensity is equal to I_0. The destructive interference occurs when $\sin^2(f/2) = 1$ or $\phi = (2k+1)\pi$ and the minimum intensity is $I_0/(1+F)$. It may be noted that in the case of multiple beam interference minimum intensity is not equal to zero. For reflected beam conditions of minima and maxima get reversed which means that two fringe systems are complementary to each other. Thus on illuminating the film with an extended source, one observes in the focal planes of lenses L_1 and L_2 a system of circular fringes with their center at the point corresponding to the wavefront emerging normally from the film (see Fig. 8.5).

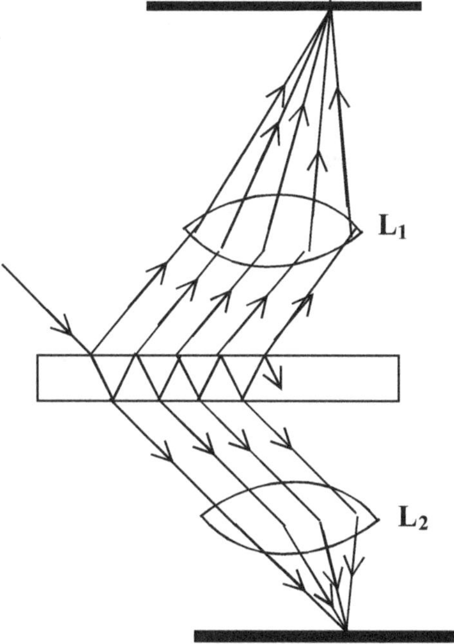

Figure 8.5: Interference phenomenon with multiple beam for reflected and transmitted beams.

In Fig. 8.6 we have plotted normalized transmitted intensity as a function of ϕ for different values of F. It may be noted from the figure that as the value of F and hence the reflectivity of the film increases the sharpness of the fringes increases which results in the increase in the resolving power of the interferometer based on multiple beam interference.

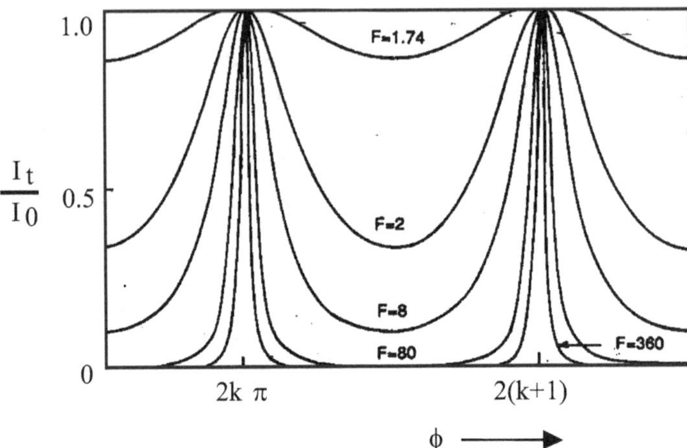

Figure 8.6: Normalized transmitted intensity as a function of phase difference for different values of coefficient of finesse.

8.3 Fiber optic interferometers

In this section we shall describe few fiber optic interferometers based on two beam and multiple beam interference.

8.3.1 Mach-Zehnder interferometer

In this interferometer (see Fig. 8.7), laser light is split into two beams of equal intensities and are injected into two single mode fibers. This can be done using a 3dB fiber coupler. One fiber acts as a sensing fiber while the other acts as a reference fiber. These two light beams are recombined using another 3 dB fiber coupler. The combined beam is then detected and the phase shift is measured. The phase shift occurs due to the change in either of the following properties of the sensing fiber:

(a) Total physical length,
(b) The refractive index, and
(c) The transverse dimensions of the fiber.

The total physical length of an optical fiber may change due to the application of a longitudinal strain or thermal expansion or due to the application of a hydrostatic pressure resulting in expansion (Poisson's ratio). The refractive index of the fiber may change due to temperature or pressure and longitudinal strain. Finally, the transverse dimensions of the fiber can change due to radial strain or longitudinal strain through Poisson's ratio or thermal expansion.

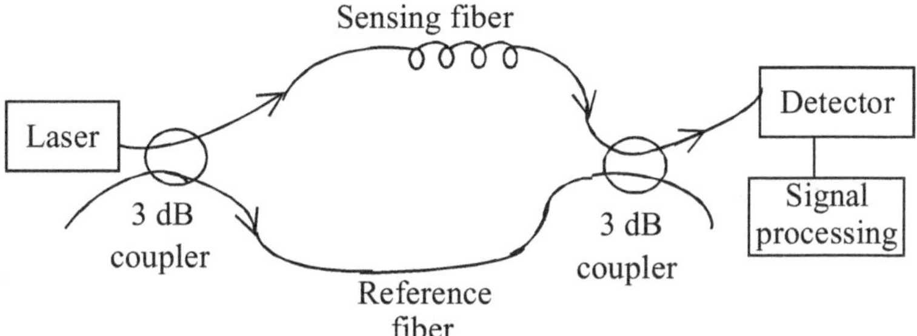

Figure 8.7: Fiber-optic Mach-Zehnder interferometer.

If the total length of the sensing and reference fibers are exactly the same or differ by an even multiple of $\lambda/2$, the recombined beams are exactly in phase and the beam intensity is maximum. However, if the difference in total lengths is odd multiple of $\lambda/2$ then the intensity of the recombined beam will be minimum. Thus a 100% modulation occurs over a change of $\lambda/2$ in the fiber length. The interferometer is so sensitive that a change as small as 10^{-13} meter in length can be detected.

8.3.2 Michelson interferometer

Figure 8.8 shows a fiber-optic Michelson interferometer. In this configuration a single 3 dB coupler is used. The coupler divides and injects light into two fibers. The ends of these fibers are reflecting. Thus the beams in both the fibers reflect and return to the original coupler and then to the detector. Thus the measurand acts on the sensing fiber twice before the recombination of the beams. The advantage of

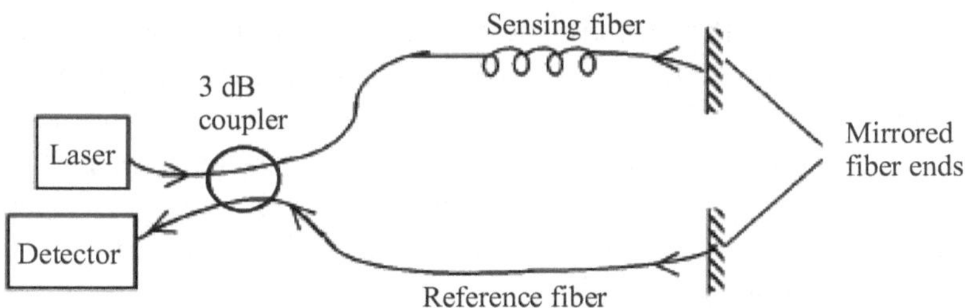

Figure 8.8: Fiber-optic Michelson interferometer.

Michelson interferometer is that it eliminates one 3 dB coupler while its disadvantage is that the coupler feeds light into both the detector and the source after reflection. This introduces source noise.

8.3.3 *Sagnac interferometer*

Another two beam interferometer, Sagnac interferometer, is shown in Fig. 8.9. In this configuration, the two beams travel through the same closed path but in opposite directions. When the loop is stationary, no phase shift occurs because the clockwise and anti-clockwise distances traveled by the light in the loop are the same. If the fiber loop is rotated about an axis perpendicular to the plane of the loop, a phase shift occurs between the two light beams traveling in opposite directions. The phase shift depends on the rotation rate, the number of turns in the loop and the area enclosed by the loop. In this interferometer change in length or refractive index is not required. The interferometer is used to sense rotation.

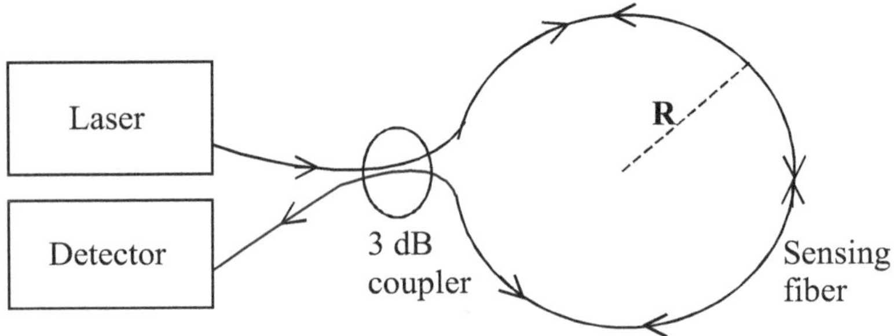

Figure 8.9: Fiber-optic Sagnac interferometer.

8.3.4 *Fabry-Perot interferometer*

The Fabry-Perot interferometer is based on the multiple beam interference. It does not require a reference fiber. The interference results from successive reflections of the initial beam. The configuration of the interferometer is shown in Fig. 8.10. It consists of two partially transmitting mirrors (~ 5% transmission). Light from the laser is injected into the fiber through the mirror 1. It is partially reflected from the mirror 2. Rest of the light is transmitted. The reflected light again falls on the mirror1 and gets partially reflected. This process continues and the intensity of the light goes on decreasing after each reflection. The beams transmitted through mirror 2 interfere. The multiple reflections increase the phase difference that results in extremely high sensitivity. This is because the sensing fiber is between the mirrors and the measurand acts on the light on every pass.

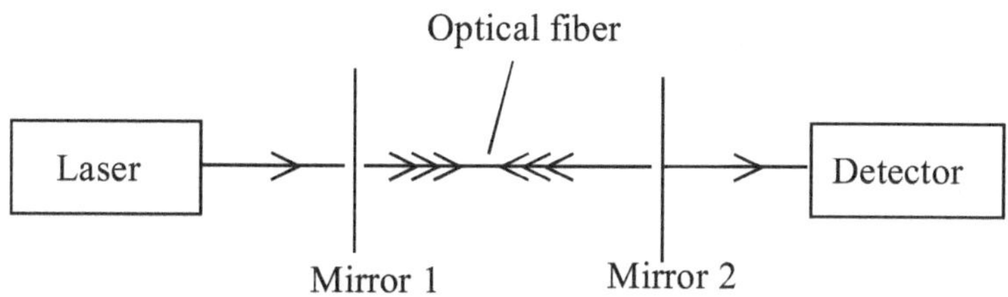

Figure 8.10: Fiber-optic Fabry-Perot interferometer.

These interferometers are applied to sense various physical and chemical quantities. In this chapter we shall describe fiber optic interferometric sensors for the detection of magnetic and electric fields, acoustics, temperature, rotation *etc.*

8.4 Magnetic field / Electric current sensor

There are a large number of fiber optic magnetic field/ electric current sensors reported in the literature. We shall describe few fiber-optic interferometric current sensors utilizing different approaches. In the first case the length of the sensing fiber is increased on heating it. The heating is done by means of flowing an electric current through the metallic coating on the fiber[1]. The increase in the length of the sensing fiber changes the optical path or the phase. Figure 8.11 shows a single mode fiber coated with a 2 μm layer of aluminium. Electrical contacts are made at both ends of the coated region (typically 10 cm long and having a 3 Ω resistance). The current to be detected is passed directly through the metallic coating of the fiber. The phase shift resulting due to heat and hence the current flown is measured by the

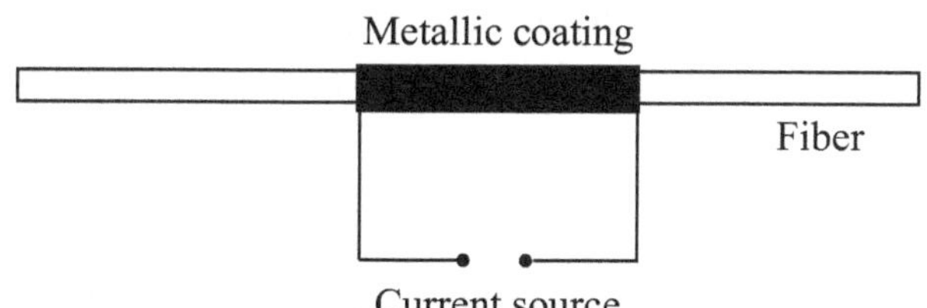

Figure 8.11: Design of the sensing arm for current measurement.

Mach-Zehnder interferometer. The sensor's dynamic range is between 1 and 10^{-9} A per meter of the fiber.

The phase change due to temperature change in the above design is proportional to the square of the electric current to be measured. This nonlinear measurement leads to a problem of low sensitivity when the electric current is weak. To enhance the sensitivity, instead of one, two metal-coated single mode optical fibers are used in the interferometer[2]. In this case the desired phase variation is linearly proportional to the electric current. The difficult part in this sensor is the preparation of two thin-film coatings with identical resistance.

In the second approach, a thin film of magnetostrictive material is applied directly onto the fiber cladding. Figure 8.12 shows the design of the probe. On application of a magnetic field, the magnetostrictive strain in the coating is mechanically coupled into the fiber core. This strain gives rise to a change in the optical path length along the fiber, which may be detected by using this fiber in one of the two arms of an interferometer. The magnetic field can then be determined by measuring the phase shift between the two arms of the interferometer. Since electric current produces magnetic field, it can also be detected in the same way. Several alloys such as nickel alloys, cobalt-iron alloys and metallic glasses exhibit magnetostrictive properties. Most of the magnetic field sensors rely on the use of alternating magnetic fields superimposed upon a dc bias field. These sensors have used phase-sensitive detection techniques to achieve the high sensitivity. In order to characterize the performance of the fiber optic magnetic field sensor, it is essential to measure a variety of parameters, including linearity of response to applied magnetic field, the relative sensitivity as a function of applied field frequency, the effects of dc magnetic-bias field on the device output, orientation response, and the impact of magnetic noise on the device sensitivity. The magnetostriction depends on the magnetic-bias-field. The induced oscillation of a magnetostrictively jacketed fiber in an ac magnetic field varies, in magnitude, as a function of the applied or ambient constant magnetic field. Thus to achieve the best performance of the device, the identification of the optimum bias conditions is required. Bias-field behaviour is related to the magnitude and distribution of the magnetic moments in the metallic glasses.

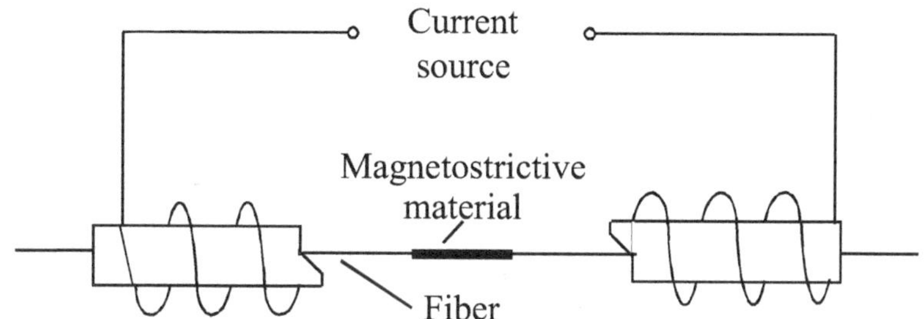

Figure 8.12: Magnetic field sensor probe based on magnetostrictive material.

The sensitivity of the sensor can also be increased if a proper composition of metallic glasses is used. But it is difficult to deposit a uniform film of metallic glass on the fiber. In addition to this, all metallic or metallic glass materials exhibit a saturation of the magnetostriction at comparatively low magnetic fields to about 100 A/m. Due to this, the sensing element may be undesirably sensitive to small dc field fluctuations and often exhibits hysteresis behaviour. To overcome this, operation of the magnetometer in a closed-loop mode is required. For this additional equipments are needed which means an increase in the cost of the sensor.

If one wants to use the sensor in an open-loop mode then a highly magnetostrictive material having high coercive field value with a reversible response to a small magnetic field should be used. Ferrimagnetic oxides with a spinel-type structure are such materials. Sedlar *et al*[3] coated single mode fibers with several magnetostrictive ceramics (magnetite, maghemite, nickel ferrite and cobalt-doped nickel ferrite) thin films. The sol-gel method is used to prepare the film. Typical coated fiber length ranges from 20 to 30 cm while the film thickness achieved is 2 μm. Measurements are carried out using Mach-Zehnder interferometer. It can detect phase shift as small as 10^{-6} rad. A He-Ne laser is used to launch light in both the arms of the interferometer (Fig. 8.13). The light exiting from the fibers are collimated and then are recombined with a beam splitter. The coated fiber is housed in two solenoid coils generating ac and dc components of the applied external magnetic field. The ac field is superimposed on a dc bias-field. The amplitude of the periodic phase shift induced in the interferometer as a result of the application of the ac magnetic field is determined using a lock-in amplifier. A linear dependence of the interferometric signal in terms of phase shift on a dc magnetic field is observed for all materials studied. No hysteresis effects are observed. Further, no saturation is seen up to dc field as high as 13.4 kA/m. Increase in ac field increases the slope of the dc magnetic response (see Fig.8.14). Figure 8.15 shows the ac field dependent phase shift for fibers coated with three different materials at a fixed dc field of 13.4 kA/m. The fiber coated with cobalt-doped nickel ferrite has maximum sensitivity. For an optical fiber having 2 μm thick coating of this material along 1 m length of the fiber the sensor can detect field as low as 2.5×10^{-3} A/m.

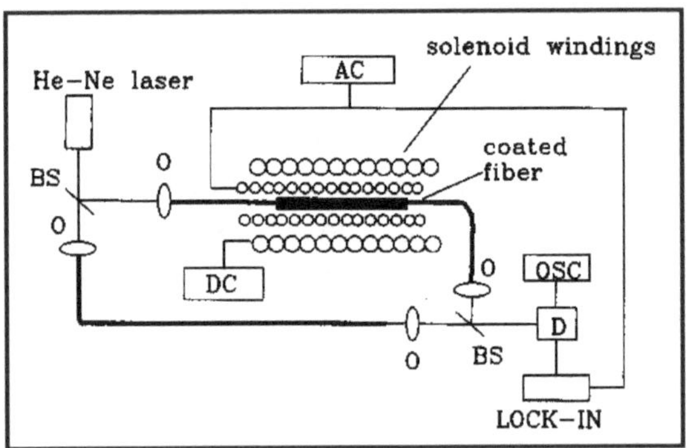

Figure 8.13: Experimental set up of the fiber optic Mach-Zehnder magnetometer. The inner solenoid provides the high-frequency signal while the outer solenoid is used to bias the ceramic-coated optical fiber to reach higher sensitivity. BS : beam splitter; O : objective; OSC : oscilloscope. Reprinted from ref. 3 with permission from OSA.

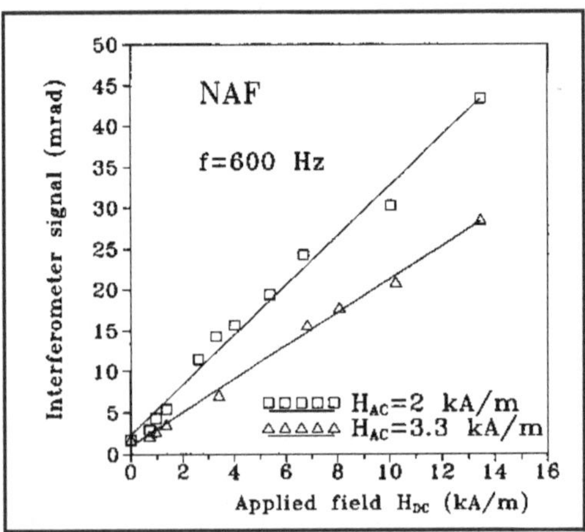

Figure 8.14: The dc magnetic field response of maghemite coated fiber. Reprinted from ref. 3 with permission from OSA.

Another approach to detect magnetic field is to wrap the fiber around a bulk magnetostrictive cylinder or tube and detect the resulting phase change.

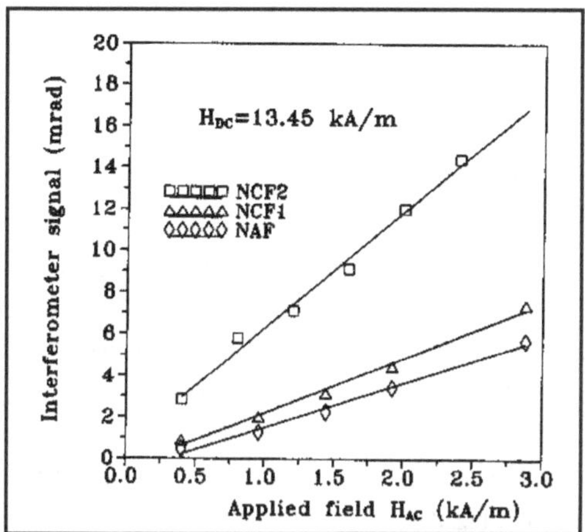

Figure 8.15: The ac magnetic field response for three different materials at a fixed dc bias of 13.45 kA/m. Reprinted from ref. 3 with permission from OSA.

Figure 8.16 shows the configuration in which the metallic glass is wrapped around a current-carrying copper cylinder. Local magnetic fields are generated by electrical currents passing through the cylinder. The fiber is bonded to the metallic glass to provide a magnetically active path length.

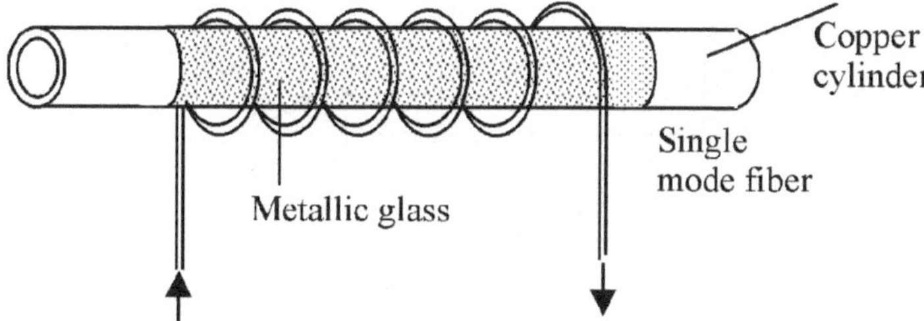

Figure 8.16: Magnetic field sensor probe with the fiber wrapped around the copper cylinder having magnetstrictive coating.

The sensors described above rely on the use of ac magnetic fields superimposed upon a dc bias field and have used phase-sensitive detection techniques to achieve high sensitivities. DC magnetic field can also be measured using a fiber coated with an amorphous $Fe_{80}B_{20}$ alloy in one arm of a modified Mach-Zehnder interferometer[4]. The alloy is a metallic glass and has magnetostrictive extension coefficient 2 orders of magnitude greater than the nickel that has been commonly used. To fabricate the probe, the fiber is stripped of their primary and secondary coatings for approximately 5 cm of its length near the center and then sputter coated with a 5 μm layer of the alloy directly onto the fiber cladding. The coated fiber along with the reference fiber is placed inside the solenoid used for generating the test magnetic fields. To minimize the differential thermal and mechanical perturbations, two fibers follow the same optical path as far as possible. In the modified interferometer two fiber ends at the output are arranged to be parallel and in close contact Fig.8.17. The lights from two fibers interfere and produce fringes. The fringe pattern is monitored by a linear array of 128 photodiodes interfaced to a microcomputer. The phase shift is determined by subtracting the normalized nonzero-field fringe pattern from a previously stored zero-field pattern. The set up eliminates noise that is due to the polarization fluctuations within the fibers. The sensor has been shown to measure a dc magnetic field in the absence of a bias field.

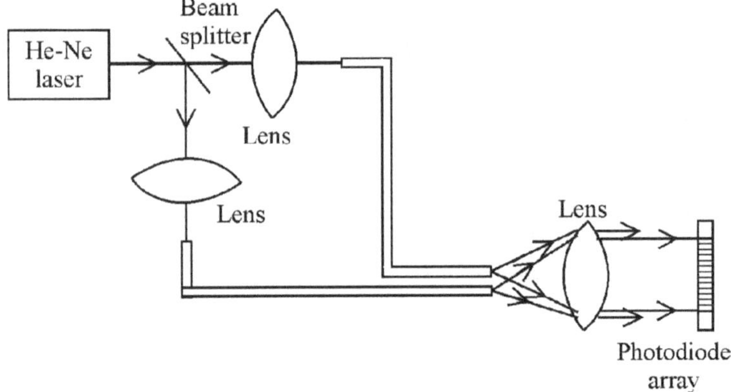

Figure 8.17: Modified fiber-optic Mach-Zehnder interferometer for the measurement of magnetic field.

8.5 Electric field /Voltage sensor

To measure the electric field/voltage, electromechanical materials such as piezoelectric materials are used. These materials can expand and contract in the presence of an electric field. Thus if the magnetostrictive coating is replaced by a piezoelectric coating (polyvinyl fluoride) in Mach-Zehnder interferometer then the sensor can be used to measure the electric field. The fiber wrapped around a tube or rod of piezoelectric material can also be used as the sensing arm. In the experiment the sensing fiber is placed between the two electrodes and different voltages are applied between them. Thus the same sensor can also be used to sense voltage. The phase shift in these sensors is directly proportional to the applied electric field.

The disadvantage of using a piezoelectric material in a fiber-optic electric field sensor is that the noise is inversely proportional to frequency. Thus, in the case of a d.c. and low frequency electric field the noise dominates and hence the resolution is very poor. To overcome this, the sensor based on electrostrictive effect has been demonstrated[5]. Although the electrostrictive effect occurs in a variety of materials, it is usually too small to be useful in sensors. However, in certain materials (*e.g.,* Ba:PZT, PLZT, $PbMnNbO_3$) the effect is large enough to render useful sensor. In the electrostrictive ceramic, the induced strain is directly proportional to the square of the applied electric field. Figure 8.18 shows the schematic of the fiber-optic interferometer to sense d.c. and the low-frequency electric field. The sensor consists of a 1 cm x 0.75 cm x 1 mm electrostrictive ceramic (Ba : PZT) bonded to a 1 cm section of three fibers in one arm of a Mach-Zehnder fiber-optic interferometer. A 30 nm gold film on each face of the ceramic allows electrical contacts. The electric-field generated strain produces phase shift of light propagating in the sensing fiber.

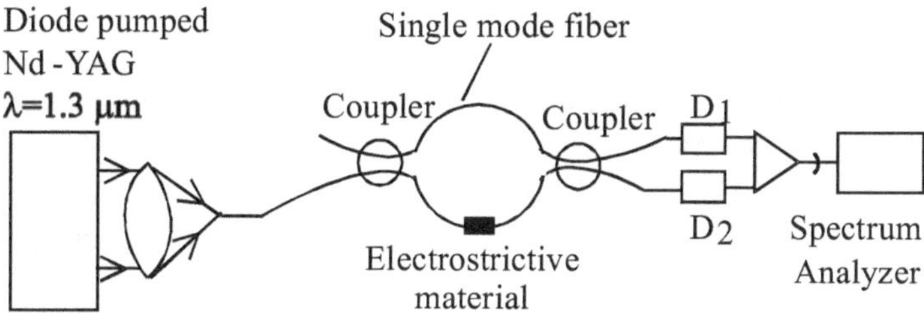

Figure 8.18: Fiber optic low-frequency and d.c. electric field sensor.

8.6 Acoustic sensor

Consider the interaction of acoustic wave with a straight length of optical fiber. There are three cases depending on the frequency range. If the acoustic wavelength is much larger than the length of the fiber then it is referred to as the hydrostatic case. If the wavelength approaches the fiber length, a length resonance is observed. In this case fiber does not respond to the axial strain. As the wavelength approaches the fiber radius, the induced radial strain becomes anisotropic. The simplest and most practical case is the hydrostatic. Typically, fiber optic acoustic sensors operate between d.c. and ~50 to 100 kHz frequency range. The sensing fiber is coated with an elastomeric material (see Fig. 8.19) which is sensitive to acoustic field. For a material with high Young's modulus, relatively thin coating is sufficient to obtain a significant increase in acoustic sensitivity over that of the bare fiber. The typical length of the fiber used is 10 m. The Mach-Zehnder interferometer configuration is used to sense the acoustic waves. To obtain the best performance, the reference fiber is desensitized using metallic coating.

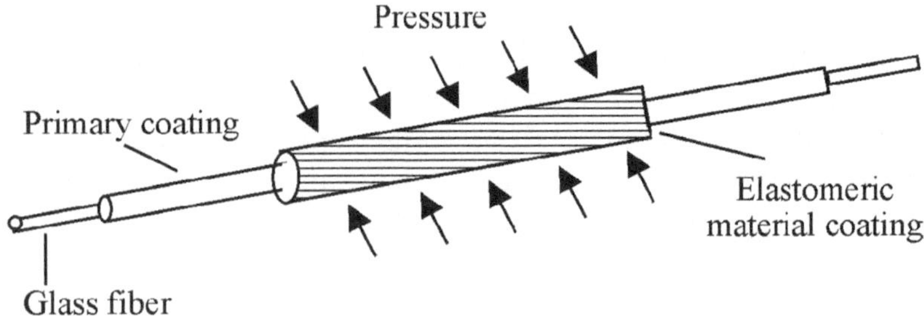

Figure 8.19: Sensing arm of a fiber optic acoustic sensor.

Another alternative configuration for the sensing arm is to use a compliant cylindrical mandrel to develop the acoustically induced strains. In the mandrel configuration, optical fiber is tightly wrapped around the compliant cylinder as shown in Fig. 8.16. It measures the radial strain of the cylinder induced by acoustic field. The strain in the optical fiber is directly related to the radial strain developed in the mandrel. If these sensors are used for underwater acoustic sensing then they are called hydrophones.

8.7 Gyroscope

The fiber gyroscope was first reported in 1976 and first appeared in civil aircraft Boeing 777 well over a decade later. Fiber optic gyroscope (FOG) is now a regular production item. It measures the inertial rotation. The advantages of fiber optic gyroscopes over the mechanical gyroscopes and ring laser gyroscopes are all solid-state, no moving parts, no warm-up time, unlimited shelf life, minimal maintenance, large dynamic range, small size and low cost. Because of these advantages several manufacturers are producing fiber-optic gyroscopes. Their scope of applications is quite broad. These are used in automobile navigation systems, pointing and tracking of satellite antenna, inertial measurement systems for commuter aircraft and missiles, as the backup guidance system for Boeing 777, mining operations, tunneling, altitude control for a radio controlled helicopter used for agricultural spraying, cleaning robots and guidance for unmanned dump trucks and carriers.

Fiber optic gyroscope is based on Sagnac interferometer described in Sec. 8.3.3. The device relies on the production of a phase difference between two counter-propagating light beams in a fiber loop as a result of system rotation. The device facilitates precise measurement of angular rate. The basic fiber gyroscope contains a multi-turn coil of fiber (Fig. 8.20). Light from a broadband source such as a light-emitting diode (LED) or a superluminescent diode is coupled into a directional coupler (DC1). The light beam then passes through a polarizer to insure the reciprocity of the counter-propagating light beams through the fiber coil. The second directional coupler (DC2) splits the two light beams into the fiber coil where they pass through a modulator that is used to generate a time varying signal for ease of processing purposes. When loop rotates the light traveling in the same direction as the rotation sees the coupler move away from it during the transit time.

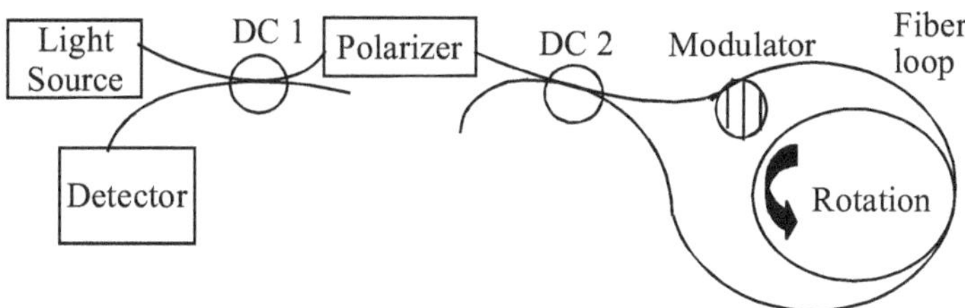

Figure 8.20: Fiber optic gyroscope.

Consequently, this stays a little longer in the loop than the light traveling against the rotation. This time difference is measured as a fringe drift at the output from the coupler. The sensitivity of the fiber gyroscope increases with the increase in the length of the fiber. It typically has a detection threshold of 0.1 to 0.01° per hour with a fiber loop area of the order of 10 cm^2. One of the main advantages of fiber gyrocope is that there are no moving parts in its design. Further its cost is less than the equivalent mechanical gyroscopes.

The ideal Sagnac interferometer is reciprocal *i.e.* the clockwise and counterclockwise optical path lengths are identical in the absence of rotation. This condition of reciprocity must be met to a very good approximation. To maintain strict reciprocity condition, the designer must use a single mode fiber with good node-stripping properties, high quality polarizers with extinction ratios of 10^5 or better and a single input-output port into the central beam splitter of the fiber coil. In addition, coherent Rayleigh backscatter from the fiber and reflections from splices and interfaces produce undesired interference terms that lead to noise far in excess of shot-noise limitations as well as additional environmental sensitivity. Thus splices and connectors in the loop must be avoided, and reflections at interfaces between fibers and other components such as integrated optic phase modulators must be suppressed. One technique for reducing these reflections is to polish the end of the fiber at an angle so that internally reflected power is outside the acceptance angle of the fiber core and is radiated into the cladding.

8.8 Temperature sensor

The temperature can be measured using fiber optic Fabry-Perot interferometer[6] shown in Fig. 8.21. The probe consists of a multimode sapphire fiber connected by fusion splicing to a single mode silica fiber. The sapphire fiber acts as the Fabry Perot cavity. The advantage of sapphire fiber is that it can tolerate high temperatures because its melting point is 2053°C.

Light from a semiconductor laser is launched into the silica fiber. A fused directional coupler is used to separate the input and output light of the sensor. The light launched into the fiber propagates to the sapphire fiber. Due to the difference in refractive index between silica and sapphire fibers, part of the incident laser light is reflected from the interface. The propagating light in the sapphire fiber is reflected at the fiber endface in air. At the silica-sapphire fiber splice, a fraction of this reflected light is recoupled into the silica single mode fiber. The reflected lights interfere and give rise to interference fringe output. The fringe contrast of the sensor output depends on the quality of the silica-to-sapphire fiber splice, more specifically on the relative position between silica and sapphire fibers at splice. An index matching liquid is used to fill the gap between the two fibers. Otherwise the gap may become another Fabry Perot cavity. The optical phase of the light travelling in the sapphire fiber is highly sensitive to temperature and longitudinal strain. Thus the sensor can be used to measure the temperature and strain. A linear relationship is reported between the fringe number change and the temperature for the temperature range 265°C to 1510°C with a resolution of 0.05°C.

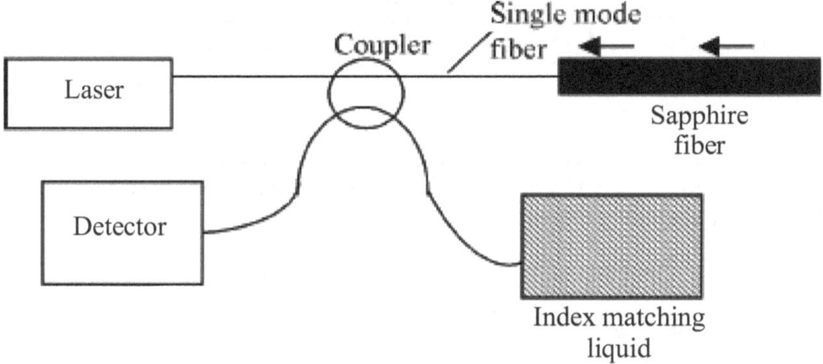

Figure 8.21: Sapphire-fiber based Fabry-Perot interferometric temperature sensor.

8.9 Hydrogen gas sensor

Hydrogen gas can also be detected using Mach-Zehnder interferometer. The sensing arm is a piece of single mode fiber with the plastic jacket stripped off from a few cm of its length. This portion of the fiber is coated with a thin film (10 μm) of palladium. When the coated portion of the fiber is exposed to hydrogen, a hydride is formed with an expanded lattice constant that stretches the optical fiber. The change in optical path length introduces the change in phase and thus the shift in fringes occurs. This kind of sensor can detect hydrogen in a concentration range 20 parts per billion to 2% in 1 atm of nitrogen[7,8].

8.10 Strain Sensor

In this section we describe a strain sensor using a thin film as low finesse fiber-optic Fabry-Perot interferometer[9]. The sensor consists of two optical fiber ends facing each other (see Fig. 8.22). The fibers are aligned using a tube and together they form an air-gap that acts as low-finesse Fabry-Perot cavity. When light is launched into the input-output

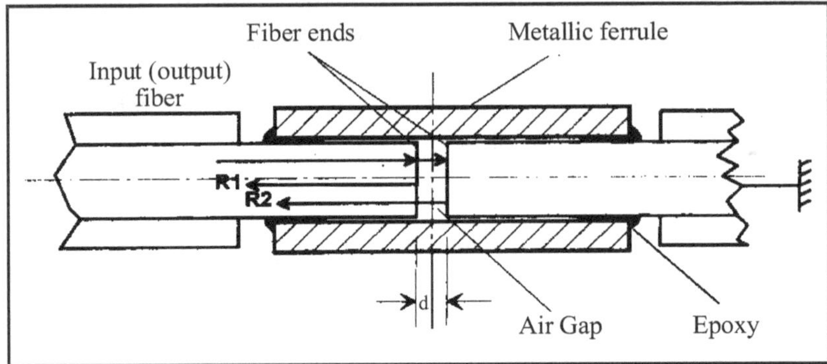

Figure 8.22: A simple fiber optic strain sensor configuration based on Fabry-Perot interferometer. Reprinted from ref. 9 with permission from Elsevier.

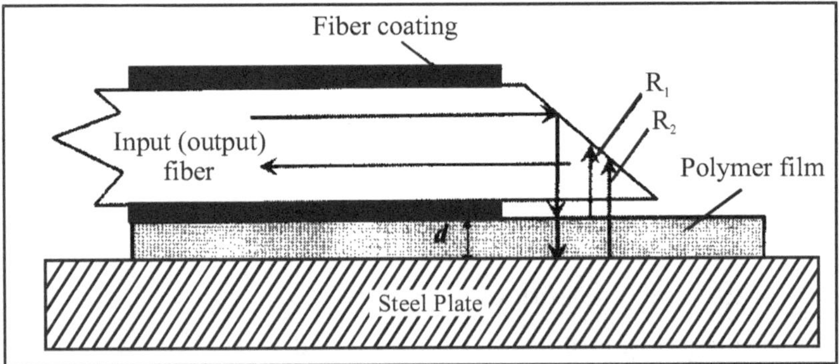

Figure 8.23: Fiber optic Fabry-Perot interferometer based strain sensor utilizing polymer film as the etalon. Reprinted from ref. 9 with permission from Elsevier.

fiber a portion of the light is reflected from each internal reflector (fiber ends). This produces an interference effect in the input-output fiber. The change in cavity length due to perturbation such as strain changes the phase and hence the relative intensity of the reflected light. Determination of change in reflected light intensity gives the amount of perturbation.

Another strain sensor that uses a thin transparent elastic polymer film as a Fabry-Perot cavity[9] is shown in Fig. 8.23. Its principle is the same as that of above. The both sides of the polymer film acts as mirrors. The light is launched in an optical fiber with one end polished at 45°. This end of the fiber is positioned close to the film as shown in figure. The polymer film is made from polyurethane that is composed of a mixture of two-phase structural domains called hard and soft segments. This provides material excellent elastomeric and high tensile strength properties. Any perturbation changing the refractive index or the thickness of the polymer film changes the intensity of the reflected light.

References

1. A. Dandridge, A.B. Tveten and T.G. Giallorenzi (1981) Interferometric current sensors using optical fibers. *Electron. Lett.* **17**, 523-525.
2. C.T. Shyu and L. Wang (1993) Electric current measurement with enhanced sensitivity using two metal-coated fibres. *J. Mod. Opt.* **40**, 2495-2499.
3. M. Sedlar, I. Paulicka and M. Sayer (1996) Optical fiber magnetic field sensors with ceramic magnetostrictive jackets. *Appl. Opt.* **35**, 5340-5344.
4. J.P. Willson and R.E. Jones (1983) Magnetostrictive fiber-optic sensor system for detecting dc magnetic fields. *Opt. Lett.* **8**, 333-335.
5. S.T. Vohra, F. Bucholtz and A.D. Kersey (1991) Fiber-optic dc and low-frequency electric field sensor. *Opt. Lett.* **16**, 1445-1447.
6. A. Wang, S. Gollapudi, R.G. May, K.A. Murphy and R.O. Claus (1992) Advances in sapphire-fiber-based intrinsic interferometric sensors. *Opt. Lett.* **21**, 544-546.
7. M.A. Butler (1984) Optical fiber hydrogen sensor. *Appl. Phys. Lett.* **45**, 1007-1008.
8. M.A. Butler and D.S. Ginley (1988) Hydrogen sensing with palladium-coated optical fibers. *J. Appl. Phys.* **64**, 3706-3711.
9. M. Jiang and E. Gerhard (2001) A simple strain sensor using a thin film as low-finesse fiber-optic Fabry-Perot interferometer. *Sensors and Actuators A* **88**, 41-46.

9

Polarimetric Sensors

9.1 Introduction

Many physical parameters such as pressure, temperature, electric current *etc*. can be measured using polarization-modulated sensors. In these sensors modulation of state of polarization of light in the sensing region by an external stimuli takes place. A variety of physical effects influence the state of polarization of light. These phenomena are Faraday effect, electro-optic effect and photo-elastic effect. Polarization modulated sensors, in some cases, offer significant advantages not only over conventional sensors but also over widely used fiber optic interferometric sensors. These sensors require control over the state of polarization of light entering the sensing region of the fiber. Thus, for these sensors, highly birefringent fibers or single mode fibers are used. In this chapter we shall first discuss magneto/electro-optic effects and then shall describe sensors for the measurement of electric current/ magnetic field, voltage, pressure, temperature *etc*. utilizing polarization modulation.

9.2 Faraday effect

The presence of magnetic field may affect the optical properties of some of the substances thereby giving rise to a number of useful devices. Faraday effect is the simplest magneto-optic effect. It concerns the change in refractive index of a material subjected to a steady magnetic field. Faraday found that when a beam of plane polarized light passes through a substance subjected to a magnetic field, its plane of polarization rotates by an angle depending on the magnetic field component parallel to the direction of propagation (see Fig. 9.1). This was discovered in 1845 by Michael Faraday. It is also called magneto-optical effect. The effect is non-reciprocal in nature. It means that when the direction of light propagation is reversed, the direction of rotation as seen from a fixed reference system, is not reversed. A light beam that passes twice through the medium in opposite directions will thus acquire a net rotation which is twice that of a single pass. The Faraday rotation is proportional to the magnetization of the material and is given by

$$\theta = \int kM.dl \tag{9.1}$$

where θ is the polarization rotation, M is the magnetization, l is the light path and k is a constant which depends on the material, wavelength and temperature. In paramagnetic and diamagnetic materials the magnetization and hence the polarization rotation is proportional to the magnetic field strength H. Thus Eq. (9.1) can be written as

$$\theta = \int VH.dl = VHl \tag{9.2}$$

where V is the Verdet constant and H is the magnetic flux parallel to the direction of propagation. In ferri- and ferro-magnetic materials the magnetization is not linearly related to the magnetic field strength, and a Verdet constant cannot be used. If a polarizer is placed at the other end of the material then the intensity of the transmitted beam is proportional to $\cos^2\theta$ or $\cos^2(VHl)$. Thus the magnetic field modulates the intensity of the transmitted beam. On the basis of this fiber optic magnetic field and electric current sensors have been developed

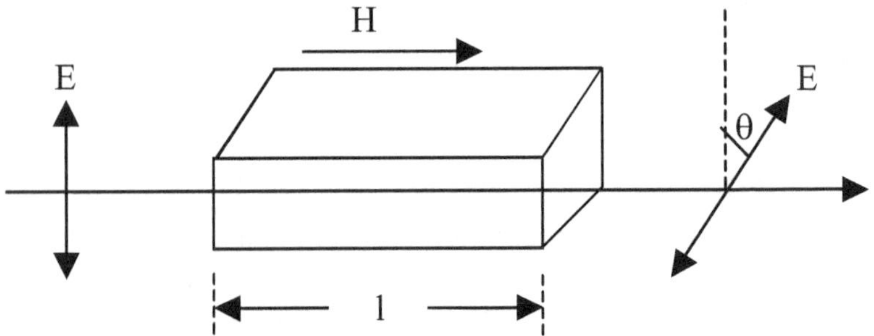

Figure 9.1: Faraday effect.

The advantage of using Faraday effect in optical fiber is that the sensor has the potential for a much longer light path than that of bulk optical material. The increase in light path increases the angle of rotation and hence the sensitivity of the sensor. Unfortunately, the Faraday effect is weak and therefore only large magnetic fields or large currents can be measured.

9.3 Kerr effect

This is the first electro-optic effect. According to this effect many isotropic substances when placed in an electric field, behave like a uniaxial crystal with the optic axis in the direction of the lines of force (or the applied field). If n is the refractive index of the substance in the absence of electric field and n_{II} and $n_{\perp}$ are the indices for the directions parallel and perpendicular to the applied electric field respectively then

$$\Delta n = n_{II} - n_{\perp} = \lambda_o K E^2 \tag{9.3}$$

where Δn is the birefringence produced, K is the Kerr constant and E is the applied electric field. From Eq. (9.3) it can be seen that the Kerr effect is proportional to the square of the field and it is therefore often referred to as the quadratic electro-optic effect. Further the effect is proportional to E^2, its sign is independent of the sign of the field. The phenomenon in liquids is attributed to a partial alignment of anisotropic molecules by the electric field. In solids the situation is more complicated. There the birefringence is proportional to the electric field and the effect is called Pockels effect. The drawback of this effect is that the large electric field is needed to see appreciable birefringence.

Below we describe various polarization modulated fiber optic sensors.

9.4 Electric current sensor

To measure electric current or magnetic field either magneto-strictive perturbation of optical fiber or Faraday effect are used. The sensors utilizing former effect have been discussed in (See 8.4; Page 169). The sensors utilizing Faraday effect that is a change of the polarization state of propagating light under the influence of a magnetic field will be described here. Faraday effect can be utilized in both bulk optical elements, planar waveguides and in optical fibers. In contrast, magnetostrictive principle requires the use of optical fibers. The Mach-Zehnder interferometer with magnetostrictive coating on the fiber of one of the arms of the interferometer is a very sensitive device for current measurement. However, it is unfortunately also sensitive to all other parameters that influence the optical length of the coated fiber *e.g.* temperature.

The basic principle of the optical fiber current sensor utilizing Faraday effect is illustrated in Fig. 9.2. A linearly polarized laser beam is launched into a single mode fiber that loops one or more times around a current carrying conductor. The light exiting from the fiber's other end is incident on a Wollaston prism. The Wollaston prism resolves the light into two orthogonal linearly polarized components. Each of these components is separately detected with photodiodes. The output signals from these photodiodes are fed into an analogue electronic circuit which computes the output function

$$S = \frac{I_1 - I_2}{I_1 + I_2} \tag{9.4}$$

where I_1 and I_2 are the intensities of the two beams. The Wollaston prism is aligned in such a way that in the absence of current the output function (S) is equal to zero. The Faraday rotation, θ, is related to the output, S, through the following relation

$$S = \sin 2\theta \tag{9.5}$$

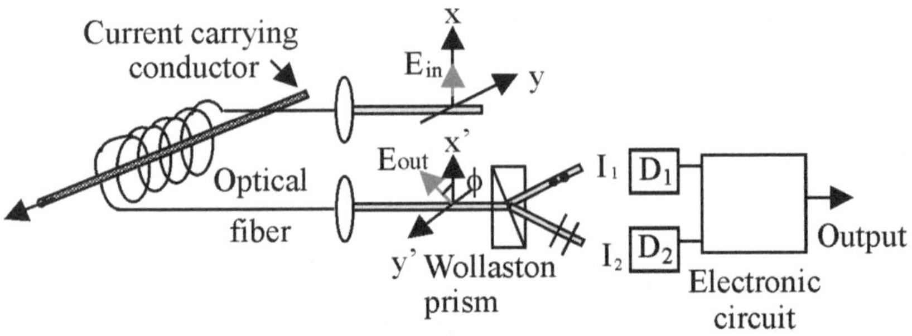

Figure 9.2: Optical fiber electric current sensor based on Faraday rotation.

When the Faraday rotation is relatively small, the output signal is directly proportional to the value of the rotation of the linearly polarized light.

If a current flows through the conductor then the magnetic field produced by the current rotates the plane of polarization of the light *via* the action of the Faraday magneto optic effect. The magnitude of rotation is proportional to the current and also the number of fiber loops around the conductor (n). Rotation of plane of polarization changes the two intensities I_1 and I_2. By measuring these two intensities and hence S the angle of rotation can be determined from Eq. (9.5). The magnetic field can then be determined from Eq. (9.2). The substitution of angle, number of fiber loops and the Verdet constant (V) of the fiber material in the following equation give the electric current

$$i = \frac{\theta}{Vn} \tag{9.6}$$

The device can measure both AC and DC currents. With 10 m fiber length coiled with a 3 cm radius, Rashleigh and Ulrich[1] obtained 0.25 mrad/A polarization rotation giving operating range 0.2 to 2000A.

The He-Ne laser operates at 632.8 nm has been used as the light source in some of the experimental devices. It has some disadvantages such as bulky, vibration sensitive, requires a high operating voltage and has relatively short life. It can, therefore, be replaced by a semiconductor laser. Although it also has some disadvantages. It operates at a longer wavelength which reduces the magnitude of the Faraday effect; it is less collimated which reduces the coupling efficiency and it has temperature dependent spectral output which requires temperature stabilization. Further semiconductor lasers are expensive. As far as fiber is concerned for the device, the single mode fiber used should maintain a linear polarization state of the light throughout its length. The change should occur in direct response to a longitudinal magnetic field. In practice this does not happen. All single mode fibers are found to be birefringent owing to slight deviations from complete circular symmetry in the core geometry, in either refractive index or diameter or both. Due to this the linear polarization state is converted into an elliptical polarization. If there is a linear birefringence effect δ due to the sensing element itself (*i.e.* the fiber), then the output signal from the detection system becomes

$$S = 2\theta \frac{\sin\delta}{\delta} \qquad \text{if } \delta >> 2\theta$$

and

$$S = \sin 2\theta \qquad \text{if } \delta << 2\theta$$

The presence of birefringence in the fiber reduces the sensitivity of the sensor as well as the stability of the output signal. Further any twist of the fiber should be avoided by holding the ends of the fiber firmly. Otherwise, it may also introduce birefringence.

A promising application for fiber optic current sensors is in the monitoring of high voltage power networks. It has been realized that optical methods for measuring current in high voltage power lines have significant advantages over conventional systems using copper wound current transformers. At many places bulk optic current transducers are currently being used which have better performance and safety characteristics than the conventional current transformers. For low voltages their prices increase. The all fiber current sensor has significant performance and cost advantages over bulk optic current transducers and conventional current transformers and can respond to almost any voltage level with high accuracy. A significant advantage is that fiber based sensors are non-conducting which eliminates high voltage insulation problems.

9.5 Magnetic field sensor

In previous chapters we have described magnetic field sensors utilizing phase modulations. Here we shall describe a magnetic field sensor that utilizes a single monomode fiber wound under tension onto a magnetostrictive cylinder such as nickel. In such configuration, magnetic field changes the state of polarization of the light propagating through the fiber coil. This occurs due to the induction of birefringence in the fiber. Figure 9.3 shows the schematic arrangement of the sensor[2]. Linearly polarized light from He-Ne laser is focused on to the face of the fiber. The other end of the fiber is coupled to a detection system similar to that in Fig. 9.2. The cylinder is placed in the magnetic field with its direction along the axis of the cylinder. The magnetic field, as mentioned above, changes the state of polarization of the light in the fiber and hence the intensities of the two beams detected by the detectors. By measuring these intensities the rotation of polarization and hence the magnetic field can be measured. The sensitivity of the sensor reported is 1.76×10^{-2} rad/m Oe and can detect magnetic field as low as 4.4×10^{-6} Oe/m of the fiber.

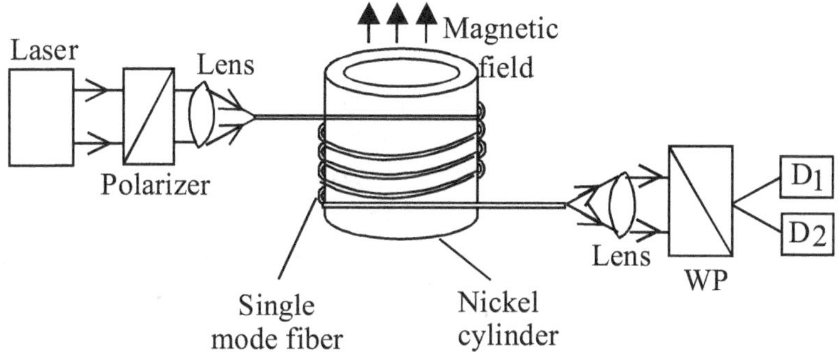

Figure 9.3: Schematic of a fiber optic magnetic field sensor based on polarization modulation[2].

9.6 Voltage sensor

For the measurement of voltage piezoelectric effect along with single mode optical fiber has been used. The Kerr electro-optic effect with optical fiber could not be used because the SiO_2 (the material of the core) has small Kerr coefficient (3×10^{-22} m/V). Kerr effect can be used if the SiO_2 core is replaced by a liquid that has high value of Kerr coefficient. Figure 9.4 shows a voltage sensor that uses liquid core optical fiber[3]. The core of the fiber is filled with a Kerr liquid, nitrobenzene. Its refractive index is 1.553 and has Kerr coefficient equal to 2.43×10^{-18} (m/V) much higher than SiO_2. The hollow fiber is made of SiO_2 with inner and outer diameters equal to 60 μm and 150 μm respectively. The refractive index of SiO_2 is 1.448 which is less than that of the Kerr liquid and hence the condition of the light guidance is satisfied. Light from a He-Ne laser is passed through a $\lambda/4$ plate to obtain a circularly polarized light. It is then passed through a polarizer to obtain the optional incident polarization angle of the linearly polarized light. Since the fiber is multimode the transmitted plane polarized light will convert into an elliptically polarized light. The degree of polarization is given by

$$P = \frac{I_{max} - I_{min}}{I_{max} + I_{min}}$$

where I_{max} and I_{min} are the semi-major and semi-minor axis lengths of the output elliptically polarized light. For Kerr effect the linear polarization should be maintained. The output state of polarization depends on the input conditions *i.e.* the convergence angle and the ratio of the spot diameter to core diameter. The electric field is applied across the fiber using the parallel plate electrodes. The electric field induces the birefringence in the fiber and hence the plane of polarization is rotated. By knowing the rotation, voltage can be measured. Since it is difficult to get perfectly linearly polarized light output the best thing is to measure the intensity transmitted through the analyzer as a function of voltage applied using a detector. The disadvantage is that the vibrations affect the state of polarization and hence the measurement should be performed carefully. The intensity of the transmitted light has been shown to vary linearly with applied voltage in the range 23 to 130 kV/m in fields.

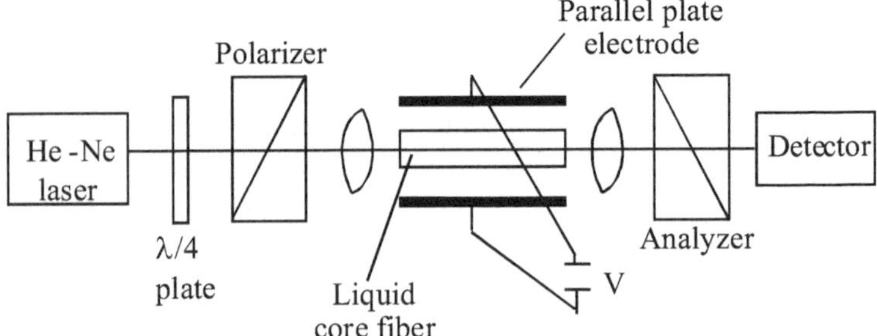

Figure 9.4: Schematic diagram of a voltage sensor based on liquid core optical fiber and Kerr effect[3].

9.7 Pressure sensor

As mentioned in the case of current sensor, the control over the state of polarization of light entering the fiber sensing region is very important. A much more useful and versatile set up would be to use a laser diode pigtailed to polarization preserving fiber. In this arrangement, polarized light from a laser diode is focused by means of a high efficiency lensing system and into a highly birefringent polarization maintaining fiber with the plane of polarization aligned parallel to the fast or slow axis. The result is a polarization preserving pigtail with a very high polarization ratio (~100:1) at the output end of the fiber. To measure pressure or temperature highly birefringent polarization maintaining fibers in the sensing region have been used. The stress rods and cladding in highly birefringent fiber have different Poisson's ratios. A pressure induced differential strain in the fiber introduces a modal birefringence which also depends on the material properties and thickness of the fiber coating. This causes a phase delay between two orthogonally polarized modes propagating along the highly birefringent sensing region. The phase delay causes a change in the state of polarization of light. Using an analyzer, a photodetector and an appropriate signal recovery system the phase delay can be determined to give a direct measure of the pressure. The polarization-modulated fiber optic pressure sensor is shown in Fig. 9.5. The pigtail has been fusion spliced at 45° (with respect to the birefringent axes) to the sensing polarization maintaining fiber in order to excite equally both polarization modes in the sensing fiber. The output fiber can be HiBi, single mode or low birefringence fiber.

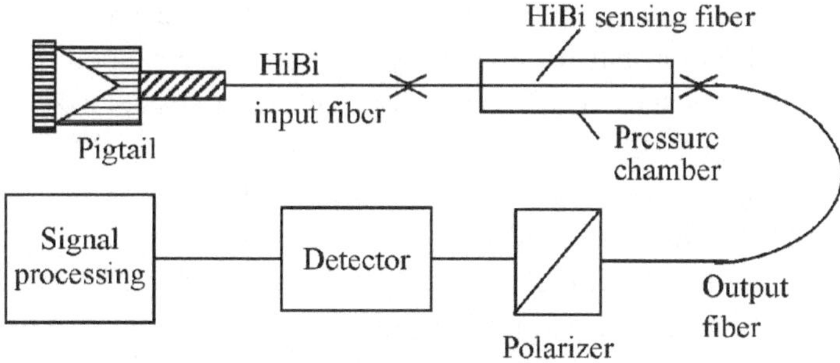

Figure 9.5: Polarization modulated fiber optic pressure sensor.

9.8 Temperature sensor

The method described above is also applicable to temperature sensor. Change in temperature induces modal birefringence in the fiber. This is because the stress rods and cladding have different coefficients of thermal expansion. Thus if the temperature of the sensing fiber changes a phase delay is induced. Since both the pressure and temperature induce modal birefringence, to measure one parameter, the other should be kept constant.

9.9 Strain sensor

The birefringence in the fiber can also be introduced by strain. Thus if a fiber is embedded in a composite material the strain can be sensed. If a polarized light is injected into the single mode fiber, it is modulated by strain-induced birefringence. As mentioned above birefringence causes the plane of polarization to rotate which causes the intensity of the transmitted light to vary. The intensity varies in a sinusoidal manner.

9.10 Immunosensors

Immunosensors based on fluorescence spectroscopy have been described in See 7.11. Here we describe an immunosensor based on the refractive index of the chemical compound. We know that the propagation velocity of light traveling through a fiber is a function of the refractive index profile surrounding the core of the fiber up to a distance approximately equal to half of a wavelength, *i.e.* the penetration depth of the evanescent field. Thus, if the change in refractive index profile takes place within this evanescent volume it can be detected as a change of the propagation velocity. Consider a D-shaped fiber with cross-section as shown in Fig. 9.6. Two orthogonallypolarized modes propagate in such fiber. Since the core of the fiber has elliptical shape, the propagation constant of light polarized along the x-axis will be different from that polarized along the y-axis; x and y are the principal axes of the ellipse. Thus if a linearly polarized light is launched into the fiber, it is transformed into elliptically polarized light at the fiber output. Further, the penetration depths of two modes are slightly different. By removing a part of the cladding from the flat side, access is made to the evanescent field region. A change in refractive index near the flat surface can now modulate the propagation constants of the two modes and hence the polarization state of the light leaving the

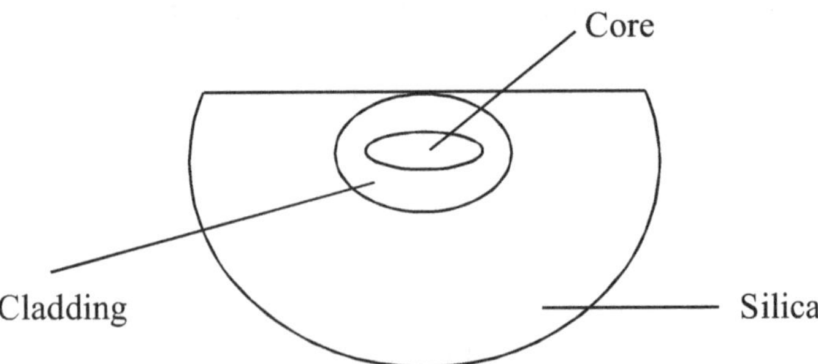

Figure 9.6: Cross-section of the D-shaped fiber with elliptical core.

fiber. This kind of sensor can also be considered as a differential fiber interferometer where both the reference and measuring arms are contained in the same fiber. Based on this principle, an optical fiber sensor for biochemical measurements has been demonstrated[4]. A few cm length of the protective silica layer and part of the cladding layer is removed from the middle portion of a fiber by wet chemical etching. This length acts as the sensing length. Light from a temperature stabilized single mode laser diode (LD) is collimated with a lens and is passed through a Faraday rotator (FR) and a half wave plate (l/2) (see Fig. 9.7).

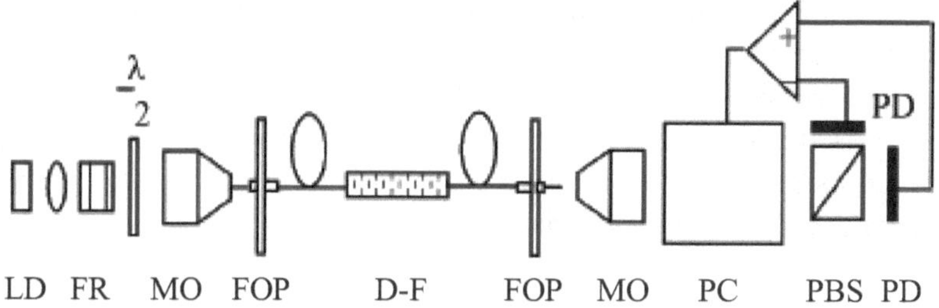

Figure 9.7: Schematic of the fiber optic biochemical sensor utilizing polarization modulation. Adapted from ref. 4.

It is then launched into the D-fiber (D-F) through a microscope objective (MO) using fiber optic positioner (FOP). The Faraday rotator prevents the light reflected from the two fiber ends and the half-wave plate to reach the laser cavity. The half wave plate launches equal amplitudes of linearly polarized light into both modes of the fiber. The sensing element is placed in a cuvette. The elliptically polarized light exiting from the fiber is collimated with microscope objective and is passed through a Pockels cell (PC). The axes of the Pockels cell are parallel to the axes of the elliptical core fiber. A voltage applied to the cell changes the state of ellipticity of the light passing through it. The light then reaches a polarizing cube beam splitter (PBS) with its axes at an angle of 45° to those of the Pockels cell. The intensities of the reflected and transmitted beams are measured with photodiodes (PD). The difference in these two intensities can be made equal to zero by applying a proper voltage to the Pockels cell. This occurs due to the transformation of elliptically polarized light into circularly polarized light. The applied voltage then becomes the signal. It varies with the process to be monitored. Initial reading is taken with the buffer solution in the cuvette. Buffer solution is then replaced by a solution containing protein. The protein gradually starts adsorbing on the flat surface of the fiber. This causes the reduction in the relative phase which changes the signal voltage with time. Thus the adsorption of antibody protein and hence its concentration can be measured.

We have described few optical fiber sensors based on polarimetric method or polarization modulation. Polarization modulation in optical fibers can also be introduced by number of other means. Mechanical twisting of a fiber induces optical activity that can also be exploited to design a sensor. Polarization modulated sensors are becoming more attractive in many ways. The simplicity of the design and the cheaper signal processing techniques make these sensors a potential rival to the widely used interferometric sensors in some areas and, like most optical fiber sensors, offer much greater sensitivity and versatility than conventional electronic/mechanical sensors.

References

1. S.C. Rashleigh and R.Ulrich (1979) Magneto-optic current sensing with birefringent fibers. *Appl. Phys. Lett.* **34**, 768-770.
2. S.C. Rashleigh (1981) Magnetic field sensing with a single-mode fiber. *Opt. Lett.* **6**, 19-21.
3. M. Kuribara and Y. Takeda (1983) Liquid-core optical fibre for voltage measurent using Kerr effect. *Electron. Lett.* **19**, 133-135.
4. R.G. Heideman, R.P.H. Kooyman and J. Greve (1993) Polarimetric optical fiber sensor for biochemical measurements. *Sensors and Actuators B.* **12**, 205-212.

10

Frequency Modulated Sensors

10.1 Introduction

There are very few frequency modulated fiber optic sensors. This is because frequency modulation of light occurs under a limited range of physical conditions. The principal one is based on Doppler effect. There are few other circumstances under which the frequency of light is modulated. These include luminescence and Raman scattering. Luminescence has been included in chapter 7 under fluorescence based sensors. In this chapter we shall focus on the sensors based on Doppler effect and Raman scattering. In particular, we shall discuss the measurement of blood flow velocity, colloidal particle motion and Raman spectrum using optical fibers.

10.2 Doppler effect

When an atom or a molecule in a low pressure gas emits radiation, a sharp quasi-monochromatic line at frequency f_o is emitted in the atom's rest frame. If the atom moves at a relative velocity v towards the observer then the observer sees a frequency

$$f_1 = \frac{f_o}{1-\frac{v}{c}} \cong f_o\left(1+\frac{v}{c}\right) \tag{10.1}$$

The shift in frequency $\Delta f = f_1 - f_o = f_o v/c$, is called Doppler shift. If the atom moves away from the observer then the frequency observed is

$$f_2 = \frac{f_o}{1+\frac{v}{c}} \cong f_o\left(1-\frac{v}{c}\right) \tag{10.2}$$

The change in frequency with velocity is called the Doppler effect. A shift of $f_0 v/c$ in frequency can also occur if a radiation of frequency f_0 is incident on a body moving with velocity v as viewed by an observer. The effect has been used to measure the blood velocity in artery and colloidal particle motion using optical fibers.

10.3 Raman effect

If a beam of light encounters a particle and if there is no exchange of energy between the radiation and the particle then the beam of same frequency will scatter in all directions. If the particle size is very much less than the wavelength of the light it is called Rayleigh scattering. The intensity of the scattered beam varies with the wavelength as $1/\lambda^4$. Higher the wavelength smaller is the scattering. If the particle size is larger than the wavelength of light then the interference between light waves scattered from different parts of the same particle takes place. This kind of scattering is called Mie scattering. Since there is no exchange of energy between the radiation and the scattering particle involved, these are called elastic scattering.

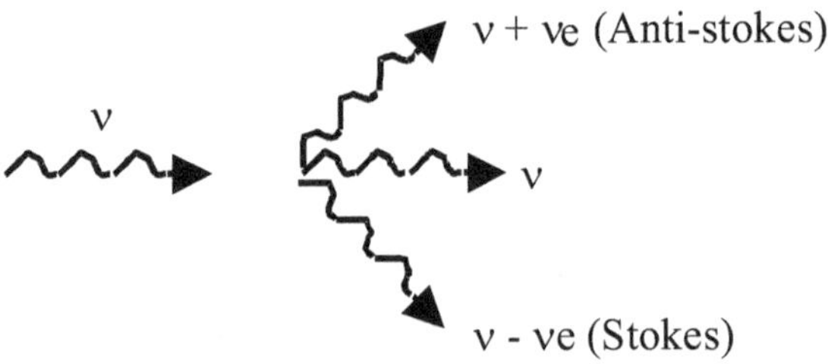

Figure 10.1: Raman scattering.

If the exchange of energy or absorption occurs then the scattering is termed as inelastic scattering. In the exchange of energy electronic, vibrational or rotational molecular energy levels of the particle are involved. If the excitation radiation is of frequency ν, the emitted radiation may have frequency larger or smaller than this (see Fig. 10.1). The frequencies so obtained are called antistokes or stokes Raman lines and the emitted radiation is called Raman scattering. The ratio of Stokes to anti-Stokes intensities is governed by the Boltzmann population of the energy levels before the excitation. At room temperature, the Stokes bands are significantly stronger than the anti-Stokes bands. Thus a typical Raman spectrum is obtained by scanning the monochromator from the exciting laser line to longer wavelength to obtain the Stokes shifted Raman spectrum. Raman lines give the information regarding the scattering particle. The Raman spectroscopy has a much lower cross-section than absorption spectroscopy, typically less by a factor of 10^9, however, because the scattered beam has a different frequency the Raman technique has many advantages over the absorption spectroscopy. The technique can be used exclusively in direct sensing. Further Raman lines are observed with virtually all organic molecules, therefore, Raman sensors are not limited to a specific class of molecules as in the case of fluorescence based sensors. The only disadvantage is that it is a weak effect and therefore sensors based on Raman effect require sophisticated instrumentation to obtain the desired signal levels at low concentrations.

10.4 Doppler effect based sensors

If we go back to (Sec. 10.2), we see that if frequency shift is known then one can find the velocity of the moving particle. There are two detection techniques for this, namely, homodyne and heterodyne. In homodyne technique, light scattered from a moving particle is optically mixed with the light scattered by other particles in the scattering volume at the detector. From this the particles motions relative to each other can be obtained. Heterodyne detection can measure both relative particle motion and the absolute velocity of particles in uniform motion. In this method, a portion of the incident light beam is separated to create a reference source of unscattered light. This unscattered light is optically mixed with the light scattered from the moving particle at the detector thus allowing the measurement of absolute velocity of the particles. Photomultiplier tube or a photodiode is generally used to detect the optically mixed light. A spectrum analyzer is used to analyze the photodetector output current and display the power or intensity spectrum of the various beat frequencies present in the signal. Both the detection techniques have been used in the fiber optic frequency modulated sensors. The advantage of using optical fiber to measure the velocity is that the system is more versatile, accepts small sample volume, can easily be transported and applicable to remote and hostile environments. The conventional systems are bulky and the sample must be placed in a special cell.

10.4.1 Measurement of blood flow velocity

The conventional methods used to measure the blood flow velocity are ultrasonic Doppler method, electromagnetic flow meter and injection method. In 1972, a new method was introduced to the medical field to determine the velocity of the blood flow[1]. The method is called Laser-Doppler Velocimetry (LDV) and was first used to measure the velocity of blood in retinal vessels. In the method laser light was focused onto the retinal vessel through the dialated pupil of the eye. Since the blood cells move in the vessel, the frequency of the light scattered by these cells was different from the incident light because of Doppler effect. The frequency shift that is proportional to the blood velocity was detected using the photon counting technique and optical mixing spectroscopy.

The first suggestion to measure blood velocity using optical fiber-based laser Doppler velocimeter[2] came in 1975. A fiber optic catheter that could be used to measure blood flow velocity was reported. The optical arrangement is shown in Fig. 10.2. Laser light, polarized perpendicular to the plane of the diagram, is incident on the beam splitter. The light reflected is focused on the input face of a multimode optical fiber using a lens. The output end of the fiber is cut and polished at an angle of 30°. This totally reflects the light out through the side of the fiber. If the catheter is inserted into the blood vessel then the light exiting from the output end of the fiber falls on the moving red blood cells (erythrocytes) that later scatter the light. The back-scattered light

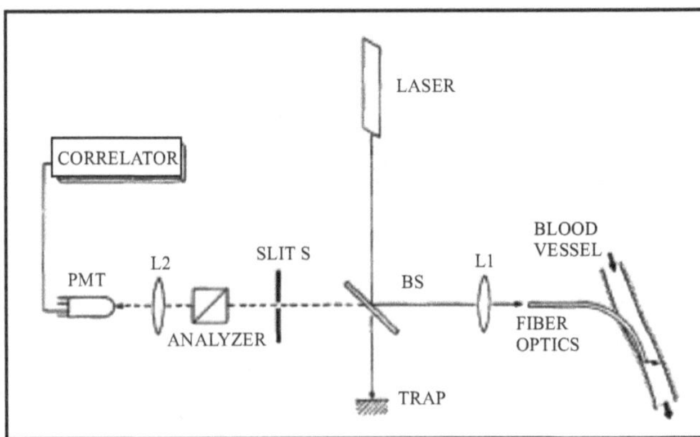

Fig. 10.2: Schematic diagram of a fiber-optic catheter. Reprinted from ref. 2 with permission from OSA.

is collected by the same fiber and is transmitted to the photomultiplier tube after passing through an analyser. Some incident light is also scattered back from the fiber exit face. This provides a reference signal. If there is a range of translational velocities of the blood cells then a spectrum of Doppler shifts is recorded using an autocorrelator. The set up can measure particle velocities in the range 0.01 to 10 cm/sec. The additional advantage of fiber optic catheter is that it is very narrow and can be easily inserted into a blood vessel without flow. Further, the blood clotting around the catheter does not occur.

A laser Doppler velocimeter for measuring skin blood flow[3] is shown in Fig. 10.3. Light from a He-Ne laser is focused onto the face of a 150 μm quartz multimode optical fiber for transmission to the skin. A 760 μm plastic multimode optical fiber is used to collect the Doppler shifted back-scattered light from the moving erythrocytes. Some of the reflected light from the stationary tissue is also collected for a reference. The total light from both sources are then passed through a laser line filter to a photodiode where heterodyne detection method is used to find the velocity of erythrocytes.

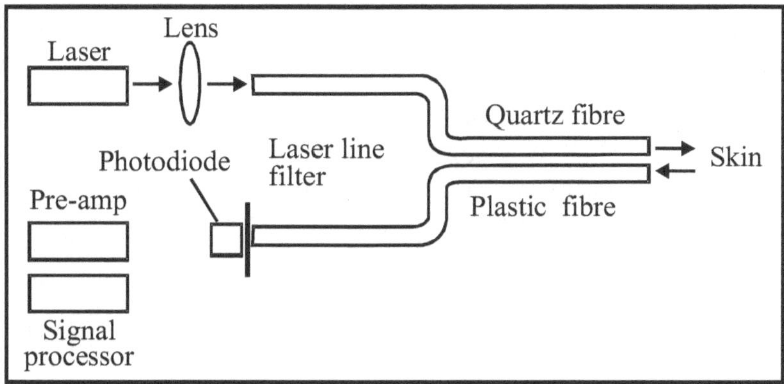

Figure 10.3: A laser Doppler velocimeter for measuring skin blood flow[3]. À© [1978] IEEE.

This apparatus has no photomultiplier tube, spectrum analyzer or autocorrelator as in Fig. 10.2 which makes the apparatus compact, portable and less expensive. Its shortcoming is the presence of noise caused by mode interference between the different longitudinal modes present in the laser cavity. Later the mode interference was suppressed significantly by using four fibers and a differential channel arrangement[4].

A high resolution single beam fiber optic laser Doppler velocimeter for the measurement of pulsatile blood flow[5] is shown in Fig. 10.4. Light from a 5 mW He-Ne laser at frequency f_o is divided into two by a beam splitter BS_1. One beam passes through a frequency shifter (acousto-optic Bragg cell) before reaching another beam splitter. The frequency shifter shifts the frequency to f_s. The second beam is transmitted through the polarization beam splitter BS_2 and focused onto the entrance of a 150 m long and 50 μm diameter graded index optical fiber. The exit tip of the fiber is inserted into the blood vessel through a hypodermic needle at a fixed angle ~ (60° optimum) to vascular flow direction. A micromanipulator allows the fiber tip to be translated laterally across the vessel diameter to give velocity profile measurements. The Doppler shifted light at frequency (f_o+Δf) is collected by the same fiber and is recombined with the reference beam which has been frequency shifted to frequency (f_o-f_s). The frequency shifting allows one to distinguish between forward and backward flow. The optical mixing is done by an avalanche photodiode. Its output is analyzed on a spectrum analyzer. The Doppler shift frequency, in this case, is given by

$$\Delta f = 2nv\cos\theta / \lambda$$

where n is the refractive index of the blood, λ is the laser wavelength in vacuum and v is the average blood velocity. The dynamic range of

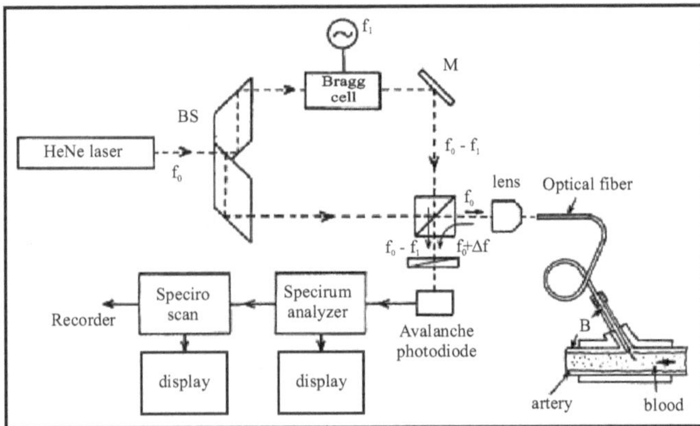

Figure 10.4: A high resolution single-beam laser Doppler velocimeter for measuring pulsatile blood flow. Reprinted from ref. 5 with permission from OSA.

the system is from 4 cm/sec to 10 m/sec with an accuracy of ±5%. Further, the measurements can be made with positional resolution of 100 μm or less and 8 msec or shorter temporal resolution. The velocimeter is capable of measuring instantaneous velocity of arterial blood flow thereby enabling physicians to monitor whether an adequate blood supply is reaching the vital organs or not.

10.4.2 Measurement of colloidal particle motion

A fiber optic Doppler anemometer for the measurement of Brownian-induced translational motion in collodial suspensions[6] is shown in Fig. 10.5. Light from a He-Ne laser is passed through a small hole in a mirror and focused onto the entrance face of a graded index multimode fiber using a lens. The exit end of the fiber is inserted into the colloidal test suspension. A portion of the light reaching exit end is reflected from the exit face of the fiber and hence provides an optical reference signal. The Doppler shifted light from the mobile colloidal particles re-enters the fiber and reaches the photodetector *via* two lenses and an analyzer. The photodiode signal of the optically mixed beams is analyzed using spectrum analyzer to know the translational diffusion coefficient of the colloidal particles. In this case heterodyne detection technique is used.

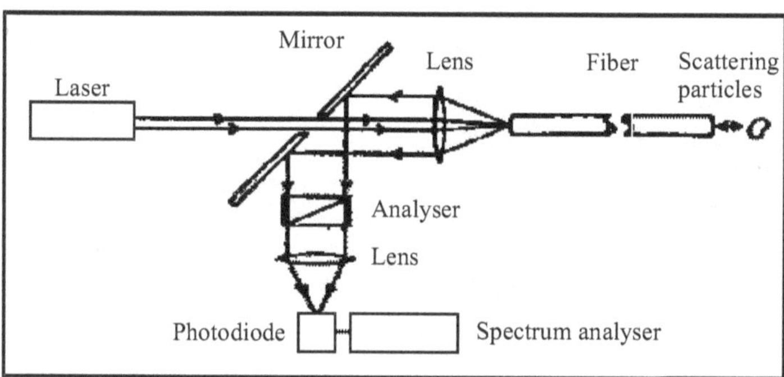

Figure 10.5: Fiber optic Doppler anemometer[6].

There are various advantages of this apparatus. Firstly, there is a very small loss of light because of hole in the mirror. Secondly, the heterodyne detection is used to achieve high sensitivity and greater dynamic range. Thirdly, the device is compact, portable and insensitive to external vibrations. The apparatus is well suited to the measurement of moving particles in a stationary medium and is used to measure Brownian motion. The movement of medium, spurious reflections from the fiber entrance face and launching lens prevent accurate measurement.

An all-fiber instrument for the measurement of concentrated colloidal and particulate dispersions[7] is shown in Fig. 10.6. Light from a He-Ne laser is focused into port1 of a graded index multimode optical fiber Y-coupler. The output fiber of the coupler (port 3) is dipped into the test sample. A small amount of light reflected from the exit tip surface again provides a reference for heterodyne detection. The back-scattered Doppler-shifted light reenters the fiber at full numerical aperture. The light coming out from port 2 reaches the photomultiplier tube after passing through a lens and a pin hole. The photomultiplier tube and a correlator are used to obtain the information about the sample. The major advantage of this system is its potential use in remote and hazardous environments or in direct industrial situations. Its disadvantage is its sensitivity to acoustic and thermal disturbances.

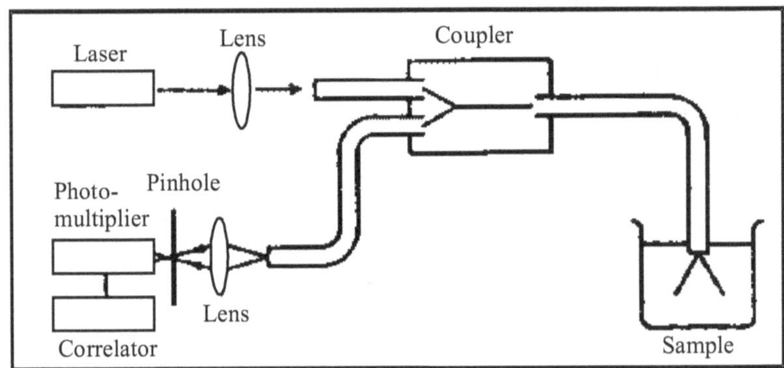

Figure 10.6: Fiber-optic Doppler anemometer using a Y-coupler. Reprinted from ref. 7 with permission from Elsevier.

Apart from these measurements laser Doppler anemometer can also be used to study the crystal growth from solution and particle size in liquid. Both single mode and multimode fibers have been used in these systems. In single mode fibers, the advantages are much smaller viewed-volume and high sensitivity. The disadvantage is that if the wavelength is changed one may have to change the fiber of different diameter to make it single mode. In the multimode fiber it is easy to couple light and they accept a greater light intensity. The intensity profile at the output end of these fibers is highly dependent on mechanical and thermal disturbances. If a sufficient amount of scattered light is available then one should prefer single mode fiber. Apart from choosing fiber there are other refinements and approaches that can make the system more compact and sensitive along with a large dynamic range.

10.5 Raman scattering based sensors

For the sensors based on absorption or reflection spectroscopy a single fiber is sufficient for both transmitting and collecting light. This is not feasible for Raman scattering based sensors. As mentioned in Sec.10.3, the intensity of the Raman scattered light is very low. Therefore, to collect a greater amount of the Raman scattered light a suitable fiber optic probe is required. Further the transmitted light excites the silica molecules of the fiber itself and thus generates Raman scattering within the fiber which interferes with the Raman scattered light collected by the fiber from the sample. Thus a multi fiber probe with the fibers performing independent functions of transmitting and collecting must be used. A single fiber surrounded by a number of collecting fibers should be used for transmitting light into the sample. The greater the number of collecting fibers in the probe, the greater the amount of the Raman scattered light that can be collected and therefore the greater the Raman signal that will be generated. However, the optimum number of collecting fibers depends on the fiber diameter and the Raman spectrometer used. The strength of the Raman signal decreases with the increase in the separation between the individual fibers. Therefore, the ends of the fibers in the probe should be in direct contact with each other. Besides minimizing the spacing between the fibers in the bundle, changing the angle formed between the axis through the end of the transmitting fiber and the axes through the ends of the collecting fibers can increase the Raman signal. For the fibers commonly used the optimum angle varies from about 6 to 12°, and the improvement in the performance varies from about 30% to 50%. To protect the fiber bundle from the environment it is enclosed in a protective shell. The environmental parameters may be pressure, temperature or some solvent. The shell should provide an impervious seal and the internal optics should not be disturbed. For many applications, stainless steel is a suitable material for the shell tube and for the shell cap. Kalrez elastomer is frequently used to seal the quartz or sapphire window into the probe body. Another type of shell that has proved useful is one entirely made from glass with the end adjacent to the bundle end being optically ground. In short, the design of the shell depends on the application.

Figure 10.7 shows a fiber optic Raman spectrometer[8]. Light from an argon ion laser is focused onto the input end of the transmitting optical fiber. The light exiting from the other end of the fiber strikes the sample. The scattered light from the sample is collected by the bundle of collecting fibers and guided back to the input slit of a monochromator. The light is dispersed by the monochromator and then imaged on the photomultiplier tube or the photodiode array.

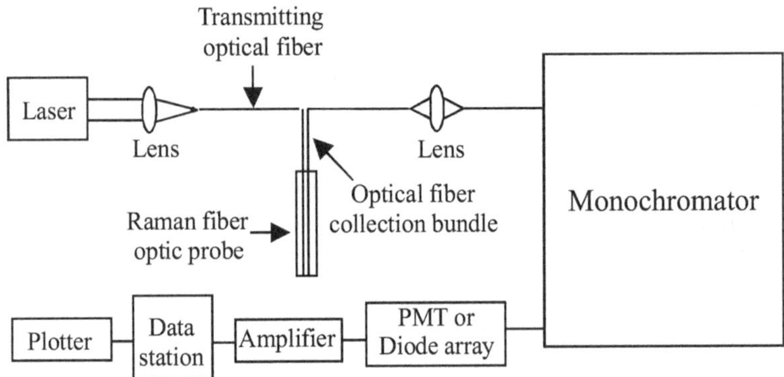

Figure 10.7: Fiber optic Raman spectrometer. Adapted from ref. 8.

The signal generated at the detector is amplified and sent to a computer to generate the Raman spectrum. The ideal Raman scattering sample is a colorless organic liquid. Spectra of such samples are obtained by inserting the probe directly into the sample container and measuring the spectrum. Raman spectroscopy can be used for the analysis of white crystalline organic materials, polymers without fluorescent additives, industrial catalysts *etc*. Raman fiber optic spectroscopy is well suited to the remote monitoring of chemical reactions. For this, the disappearance of reagent Raman bands and the appearance of intermediate or product Raman bands are monitored. On-line monitoring of the manufacturing of chemicals and materials in the chemical industries can be very well carried out by fiber optic Raman spectroscopy.

References

1. C. Riva, B. Ross and G. Benedek (1972) Laser-Doppler measurements of blood flow in capillary tubes and retinal arteries. *Invest. Ophthalmol.* **11**, 936-944.
2. T. Tanaka and G.B. Benedek (1975) Measurement of the velocity of blood flow in vivo using a fiber optic catheter and optical mixing spectroscopy. *Appl. Opt.* **14**, 189-196.
3. D. Watkins and G.A. Holloway (1978) An instrument to measure cutaneous blood flow using the Doppler shift of laser light. *IEEE Trans. Biomed. Engg*. BME-**25**, 28-33.
4. G.E. Nilsson, T. Tenland and P.A. Oberg (1980) A new instrument for continuous measurement of tissue blood flow by light beating spectroscopy. *IEEE Trans. Biomed. Engg*. BME-**27**, 12-19.
5. H. Nishihara, J. Koyama, N. Hoki, F. Kajiya, M. Hironaga and M. Kano (1982) Optical fiber laser Doppler velocimeter for high-resolution measurement of pulsatile blood flows. *Appl. Opt.* **21**, 1785-1790.
6. R.B. Dyott (1978) The fiber-optic Doppler anemometer. *Microwaves, Optics and Acoustics* **2**, 13-18.
7. H. Auweter and D. Horn (1985) Fiber-optical quasi-elastic light scattering of concentrated dispersions. *J. Colloid. and Interface Science* **105**, 399-409.
8. A.M. Leugers and R.D. McLachlan (1988) Remote analysis by fiber optic Raman spectroscopy. *Proc. SPIE* **990**, 88-95.

11

Fiber Bragg Grating Based Sensors

11.1 Introduction

In previous chapters we have described fiber optic sensors based on intensity and phase modulations. These sensors have some problems in practical applications that need to be solved. Some of these are fluctuation of light source power, loss in couplers, scattering and absorption effects that affect measurement performance, temperature drifts and vibrations. In comparison to intensity and phase modulated sensors fiber Bragg grating (FBG) sensors have following advantages:

1. High resolution
2. Insensitive to intensity fluctuation because the signal obtained from an FBG sensor is encoded directly in wavelength domain.
3. Provision of well localized sensing region
4. Linear output as a function of the measurand
5. The capability of multiplex many sensors along one fiber

These advantages improve some performance of the fiber optic sensors. However, fiber Bragg grating sensors have some limitations. FBGs are sensitive to both temperature and strain. Discrimination of their effects is required. The other problem with FBG sensor is the requirement of a demodulation technique for the measurement of wavelength shift. The measurement of Bragg wavelength shift with conventional spectrometers or monochromators is not a good solution. This is because of their bulk-optical nature, size, lack of ruggedness and limited resolution. The wavelength resolution less than 1×10^{-3} nm is often required. In the beginning this resolution was achievable only by using expensive, high resolution spectrum analyzers. This severely limited the practicality and cost effectiveness of these sensors in engineering applications. The development of several novel and practical demodulation systems in addition to low cost production of FBG in the past few years has tremendously increased the field applications of FBG sensors.

11.2 Bragg's law

When X-ray beam is incident on the regular array of atoms making up a crystal the resulting diffraction pattern consists of complex pattern of spots. In 1912, the diffraction pattern was interpreted by Nobel Laureate Sir Lawrence Bragg by means of a very simple mathematical expression. He visualized three dimensional diffraction process as a combination of partial reflections from imaginary planes containing sheets of atoms. If these regularly spaced planes could provide reflected beams with exactly the right phase delay then the individual beams would add up to produce a strong beam at the angle given by the Snell's law of reflection. If the X-ray beam of wavelength λ is incident normally on a set of planes spaced at a distance Λ from one another in a medium of refractive index n, a strongly diffracted beam will occur in exactly the reverse direction to the incident beam when

$$\lambda = 2n\Lambda \tag{11.1}$$

This is the special case of Bragg's law. The relevance of Bragg's law to the propagation of light in optical fibers was realized in 1978 and since then fiber Bragg gratings have been used in both telecommunications and sensing systems.

A fiber Bragg grating consists of a longitudinal, periodic variation in the refractive index of the core of an optical fiber as shown in Fig. 11.1. The length of the FBG, which is an integral part of the fiber, is normally few millimeters. When light from a broadband source is launched in the fiber, the FBG reflects one wavelength intensely that satisfies Eq. (11.1) while the rest of the wavelengths get transmitted. The reflected wavelength is called Bragg wavelength (λ_B) and is determined by the period of the grating (Λ). The Bragg wavelength is sensitive to changes in either the fiber-core refractive index or the period of the grating. This makes it an excellent sensor for measurement of mechanical or thermal perturbations. Unlike intensity modulated sensors which need a very stable light source and interferometric sensors which need a highly coherent laser, fiber Bragg grating based sensors generally require a broadband source and a high resolution wavelength shift detection system.

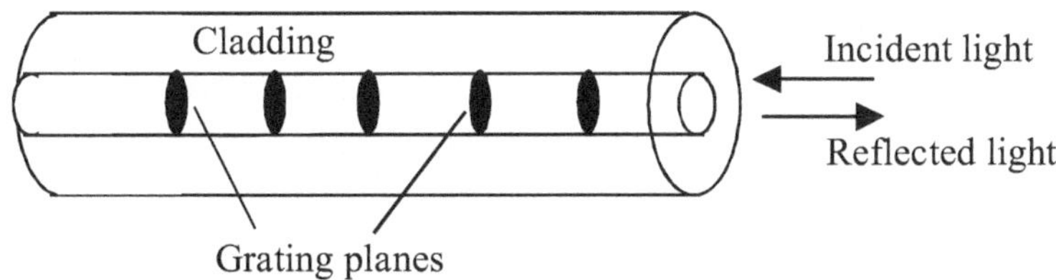

Figure 11.1: A schematic of the in-fiber grating structure of a fiber Bragg sensor

11.3 Fabrication of fiber Bragg gratings

Fiber Bragg gratings are fabricated by exposing a step index, germanosilica fiber to intense ultra violet light. The krypton fluoride (KrF) excimer laser at 248 nm or frequency doubled argon-ion laser at 244 nm are the suitable sources for this purpose. The absorption of UV light causes a permanent change in the refractive index of the fiber. There are two basic techniques that are being used to produce the required high-frequency index modulation in the fiber.

1. Interferometric method: In this method a single UV laser beam is split into two components, which are subsequently recombined to produce an interference pattern as shown in Fig. 11.2. The fiber, stripped of its protective plastic coating (which would otherwise block the UV radiation) is stretched across the interference region so that the stripes of UV light imprint the grating structure into the fiber. Photons of UV in the range of 240 to 260 nm can change the linkages of the Ge-oxygen bonds, causing permanent changes in the local refractive index. For symmetric incidence and an angle (θ) between the UV beams, the pitch of the recorded grating is given as $\Lambda = \lambda/2 \sin(\theta/2)$, where λ is the wavelength of illumination. The primary disadvantage of this technique is that the interference fringe spacing and placement is highly sensitive to the optical alignment of the system. In addition, maintaining adequate fringe contrast requires high mechanical stability and isolation from ambient vibration. In another interferometric set up a mirror positioned

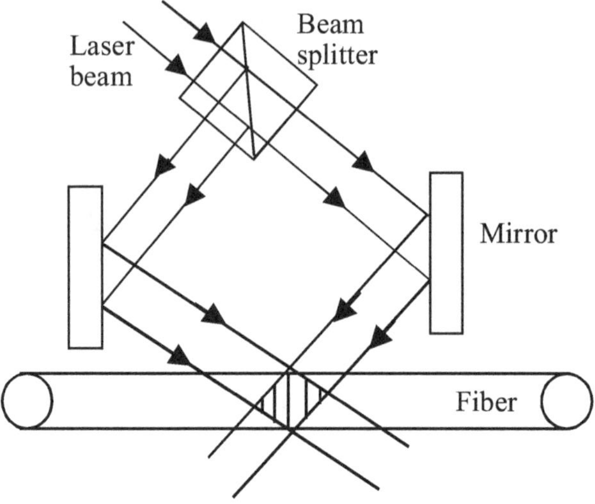

Figure 11.2: Interferometric method to produce fiber Bragg grating.

perpendicular to the fiber is used to fold half of the laser onto its other half Fig. 11.3. This is a more stable geometry and permits easier adjustment of the pitch. In both the cases a highly coherent laser source is required.

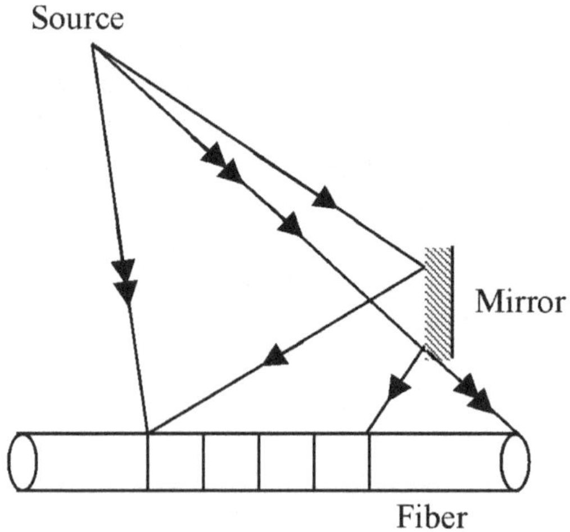

Figure 11.3: Interferometric method with single mirror to produce fiber Bragg grating.

2. ***Phase-mask method:*** In this method a diffraction grating (phase mask) is used to split a single laser beam into several diffractive orders as shown in Fig. 11.4. Interference between various orders produces the required pattern in the fiber. The fiber-phase mask assembly can be illuminated either by a large beam to cover whole of the fiber Bragg grating or by a small scanning beam. The phase-mask method is not as flexible as the interferometric method. It is far less sensitive to vibration and alignment because the fiber is kept in close proximity to the phase mask. This makes it more suitable for production environments. The performance of fiber Bragg grating greatly depends on phase mask quality. Non-uniformity in the phase mask's efficiency, pitch or other qualities can affect the grating's spectral response.

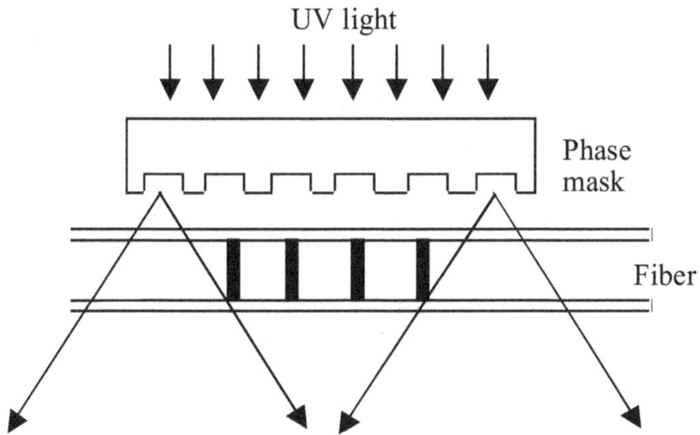

Figure 11.4: Production of fiber Bragg grating using phase mask method

11.4 Sensing principles

Fiber Bragg gratings have been reported for the measurements of strain, temperature, dynamic magnetic field *etc*. The Bragg wavelength will vary with the change of these parameters experienced by the fiber. The change in Bragg wavelength with different parameters is discussed below.

11.4.1 Measurement of strain

In the case of load induced strain, both the physical elongation of the optical fiber (and the corresponding change in the grating pitch) and the change in the refractive index of the fiber due to photoelastic effects take place. As a result the shift in Bragg wavelength occurs. The shift in Bragg wavelength with strain can be expressed as

$$\Delta\lambda_B = (1 - p_e)\lambda_B \varepsilon \qquad (11.2)$$

where ε is the applied strain and p_e is an effective photoelastic coefficient term which is given by

$$p_e = (n^2/2)[P_{12} - \mu(P_{11} + P_{12})] \qquad (11.3)$$

where P_{ij} are the Pockel's coefficients of the strain optic tensor and μ is the Poisson ratio of the optical fiber. For germanium doped silica, the effective photoelastic constant is equal to 0.22. The normalized strain response of a 1.3 μm FBG found to be

$$\frac{1}{\lambda_B}\frac{\Delta\lambda_B}{\Delta\varepsilon} = 0.78\text{x}10^{-6}\,\mu\varepsilon^{-1}$$

The above responsivity tells that 1 nm shift in Bragg wavelength occurs per 1000 με strain at 1.3 μm.

11.4.2 Measurement of temperature

The dependence of Bragg wavelength on temperature occurs due to two primary effects : the temperature dependence of the refractive index of the glass and the thermal expansion of the glass. In the case of silica fiber, the former has a dominant effect, accounting for about 95% of the observed shift in the Bragg wavelength. The shift can be expressed as

$$\Delta\lambda_B = \left[\frac{1}{\Lambda}\left(\frac{\delta\Lambda}{\delta T}\right) + \frac{1}{n}\left(\frac{\delta n}{\delta T}\right)\right]\lambda_B \Delta T \qquad (11.4)$$

where the first term relates to the thermal expansion of the fiber while the second term is due to the temperature dependence of the refractive index. For the silica fiber the change in temperature by 0.1°C changes the wavelength by 1 pm at $\lambda_B \approx 1.55$ μm. The measurement of such a small change in wavelength is more of a challenge. Various techniques have been used to measure such a small change.

11.4.3 Measurement of pressure

When the fiber is compressed the change in Bragg wavelength due to change in pressure by (ΔP) is given by

$$\Delta\lambda_B = \left[\frac{1}{\Lambda}\left(\frac{\delta\Lambda}{\delta P}\right) + \frac{1}{n}\left(\frac{\delta n}{\delta P}\right)\right]\lambda_B \Delta P \tag{11.5}$$

where the first term relates to the pressure induced expansion of the fiber while the second term is due to the pressure dependence of the refractive index of the fiber core.

11.5 Strain sensor

Figure 11.5 shows a FBG strain sensor. The FBG is connected to a broadband source via one port of a 3 dB coupler. This coupler also connects the returned signal from FBG to another wavelength dependent coupler. A fiber isolator is used to protect the source from the returned signal.

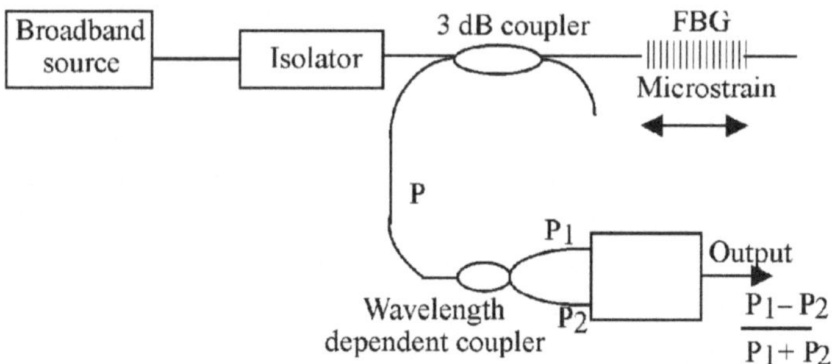

Figure 11.5: Fiber Bragg grating sensor for the detection of strain. Adapted from ref. 1.

The output intensities of both ports on the output coupler are detected. The ratio of the difference to the sum of outputs is determined. The ratio is used to eliminate the intensity variation effects. The normalized output depends on the Bragg wavelength. To calibrate the sensor, a conventional resistive strain gauge is mounted alongside the grating. The calibration curve[1] is shown in Fig. 11.6. The FBG sensor can also be used to sense vibration[2]. The FBG is attached to the vibrator which in turn produces a dynamic strain in the grating. To detect the vibration, the reflected wavelength modulated light is processed by means of a Mach-Zehnder interferometer.

11.6 Pressure sensor

The optical fiber Bragg grating sensor for the measurement of pressure[3] is shown in Fig. 11.7. Light from a 1550 nm fiber-pigtailed

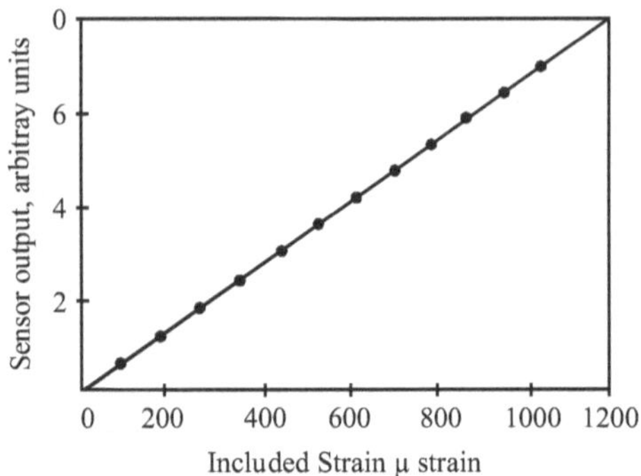

Figure 11.6: Calibration curve of the strain sensor[1]. À© [1994] IEEE.

ELED is coupled into the single mode FBG sensor head using a fiber directional coupler. The probe is installed within a cavity in a high pressure vessel. When the probe is pressurized its reflectance is monitored using an optical spectrum analyzer connected to the returned port of the coupler. The undesirable reflection from the unused output port of the coupler is suppressed using index-matching oil. The reflection from the end of the fiber with Bragg grating is suppressed by immersing it in the hydraulic oil itself. The pressure in a pressure housing capable of being used up to 70 MPa is increased using a hydraulic pump. The response curve of the FBG pressure sensor is shown in Fig. 11.8. The measured fractional change in wavelength per unit change in pressure is – 1.98×10^{-6}/MPa. If the temperature of

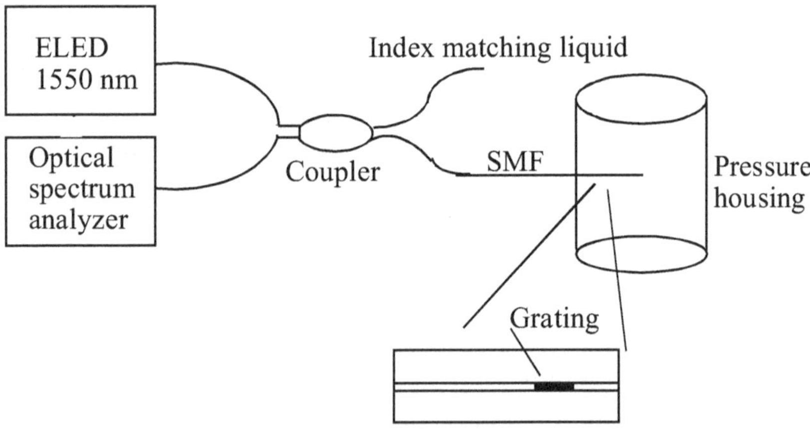

Figure 11.7: Schematic diagram of a fiber Bragg grating pressure sensor

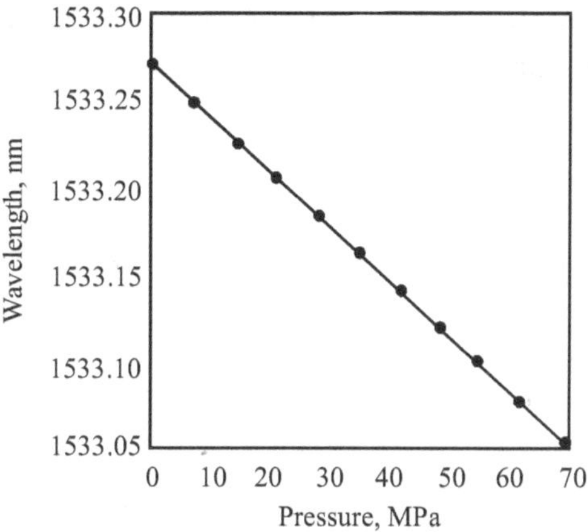

Figure 11.8: Calibration curve of the FBG pressure sensor[3]. À© [1993] IEEE.

the oil around the grating is varied then the sensor will act as temperature sensor. The temperature response of the same sensor is shown in Fig. 11.9. The measured fractional change in wavelength per unit change in temperature is $6.72 \times 10^{-6}/^{\circ}C$.

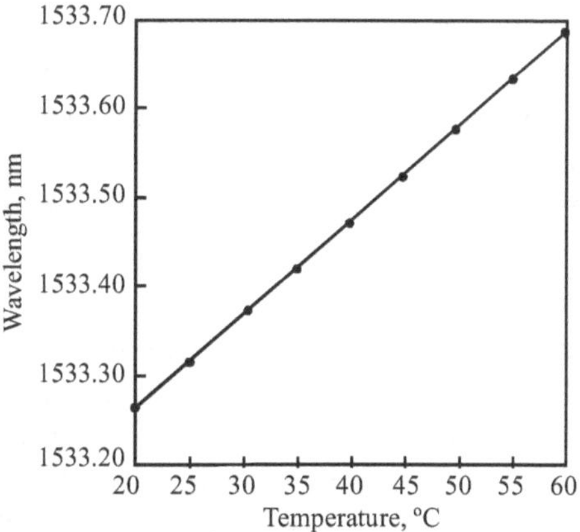

Figure 11.9: Calibration curve of the FBG temperature sensor[3]. À© [1994] IEEE.

11.7 Acceleration sensor

Fiber Bragg gratings can also be used to measure acceleration[4]. The elongation of the fiber with strain is used to sense the acceleration. The elongation increases the period of the grating and shifts the spectral response toward the longer wavelength. The schematic diagram is shown in Fig. 11.10.

The operating range of the sensor is up to 1,70, 000g; where g is the acceleration due to gravity.

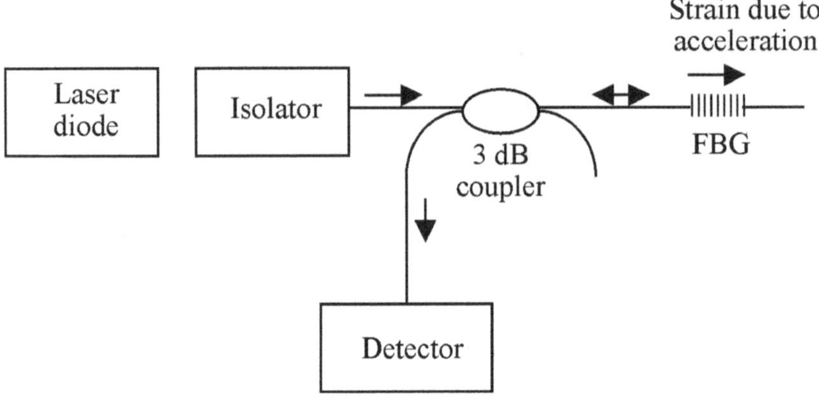

Figure 11.10: FBG sensor for acceleration[4].

11.8 Petroleum hydrocarbon detection

Petroleum hydrocarbons are hazardous substances. Their leaks in pipelines and storage tanks are very dangerous. Their leakage may lead to environmental pollution and even to catastrophic results. Due to these, their detection, localization and cure at the initial stage become important. A FBG sensor has been developed to detect hydrocarbons[5]. In the sensor FBG is attached to a special type of polymer which expands with the absorption of petroleum hydrocarbon. The swelling of polymer produces a force that acts on the fiber. This force results in an axial strain on the grating and hence changes the grating pitch that shifts the Bragg wavelength. The white butyl rubber is used as the material. This swells in hydrocarbon medium without dissolution and without significant change of other mechanical properties. The FBG sensor probe is shown in Fig. 11.11. The Bragg grating is placed within a cylinder of butyl rubber along its axis. The swelling of polymer under hydrocarbon influence affects the fiber length through a couple of metallic plates-anchors attached to the fiber on both sides of the grating. The elongation of fiber shifts Bragg wavelength towards longer wavelength. The shift of Bragg wavelength is more than 2 nm for 20 min gasoline influence[5]. This shift is more than the shift due to possible environmental temperature variation.

11.9 Humidity sensor

Fiber Bragg grating can also be used to detect humidity. The principle is based on the strain effect induced on the FBG through the swelling of the moisture sensitive polymer coated on the FBG. The choice of the polymeric material plays a crucial role in the performance of the sensor. When it is coated onto a FBG, the humidity response of the sensor monitored, in terms of Bragg wavelength shift, is dependent on how effectively the swelling of the material stretches the fiber.

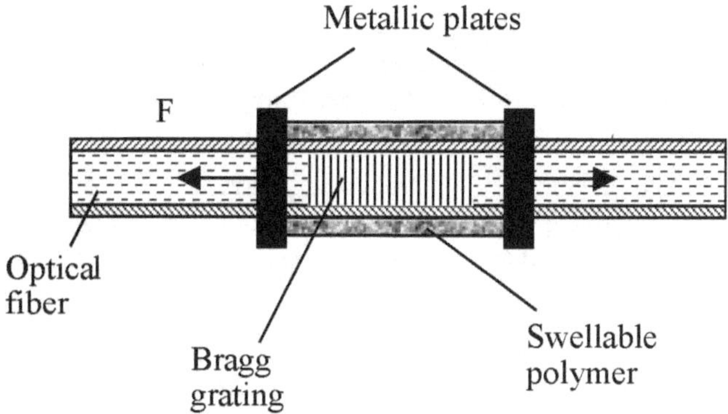

Figure 11.11: FBG sensor configuration for sensing hydrocarbons[5].

For the best performance, the material should have high moisture swelling property and the strength of the hydrated material should not be weak. In one of the humidity sensors[6], polyimide has been used. The advantage of this material is that its volume expansion when exposed to humidity change is linear and FBG sensor coated with this exhibits a linear and reversible response with humidity change. To fabricate the probe, the gratings are dip-coated in a polyimide solution. The Bragg wavelength shift increases with the increase in the humidity. The sensor response is linear from 23 to 97% RH at constant temperature. The sensitivity of the sensor increases with the increase in the thickness of the polymer film but the response time decreases. The optimum thickness is reported to be about 24 μm.

References

1. M.A. Davis and A.D. Kersey (1994) All fiber Bragg grating strain sensor demodulation technique using a wavelength dependent division coupler. *Electronics Lett.* **30**, 75-77.

2. A.D. Kersey and T.A. Berkoff (1992) Interferometric signal processing for strain and vibration sensing using two-mode and Bragg grating fiber sensors. *Proceed of the ADPA/AIAA/ASME/SPIE on Active Materials and Adaptive Structures*, IOP Publishing, Bristol, pp. 651-656.
3. M.G. Xu, L. Reekie, Y.T. Chow and J.P. Dakin (1993) Optical in-fiber grating high pressure sensor. *Electronics Lett.* **29**, 398-399.
4. S. Theriault, K.O. Hill, F. Bilodeau, D.C. Johnson and J. Albert (1996) High g-accelerometer based on an In-Fiber Bragg Grating sensor. In: *Proceed of the 11th Conference on Optical Fiber Sensors*, Sapporo, Japan, pp. 196-199.
5. V.V. Spirin, M.G. Shlyagin, S.V. Miridonov, F.J. Mendieta Jimenez and R.M. Lopez Gutierrez (2000) Fiber Bragg grating sensor for petroleum hydrocarbon leak detection. *Opt. Laser Engg.* **32**, 497-503.
6. T.L. Yeo, T. Sun, K.T.V. Grattan, D. Parry, R. Lade and B.D. Powell (2005) Characterisation of a polymer-coated fibre Bragg grating sensor for relative humidity sensing. *Sensors and Actuators* B. **110**, 148-156.

12

Surface Plasmon Resonance Based Sensors

12.1 Introduction

Among a variety of optical evanescent wave techniques, spectroscopies based on the resonant excitation of surface plasmons have matured into a widely accepted surface analytical method used in different fields. Liedberg *et al*[1] were the first who demonstrated the exploitation of surface plasmon resonance (SPR) for chemical sensing. Since then the SPR sensing principle has received much attention. In this chapter we shall first describe the principle of this technique and then few sensors that have utilized this phenomenon.

12.2 Principle of SPR

Surface plasmons are the surface localized electromagnetic waves produced by the collective resonating oscillation of free electrons on the plasma surface (for example, metal). The surface plasmon is a transverse wave propagating along the plasma surface with the oscillating electric field vector normal to the surface. Because the surface plasmon is a transverse magnetic mode, it can be excited by p-polarized light. When the wave vector and the frequency of the incident light coincide with those of the surface plasmon, this light resonantly excites the surface plasmon. The wave vector K_{sp} of the surface plasmon on the boundary between a metal and a sample at frequency ω is approximately given by[2]

$$K_{sp} = \frac{\omega}{c}\sqrt{\frac{\varepsilon_m \varepsilon_s}{\varepsilon_m + \varepsilon_s}} \qquad (12.1)$$

where ε_s and ε_m are the real parts of the dielectric constants of the sample and the metal, respectively and c is the velocity of light in vacuum. The wave vector of the light at frequency ω propagating through the sample medium is given by

$$K_s = \frac{\omega}{c}\sqrt{\varepsilon_s} \tag{12.2}$$

Since $\varepsilon_m < 0$ and $\varepsilon_s > 0$, for a given frequency, the wave vector of surface plasmon is greater than the wave vector of the light in sample medium. Therefore to excite SPR the momentum and hence the wave vector of the exciting light in sample medium should be increased. Kretschmann[3] proposed a way to excite SPR by an evanescent wave from a high refractive index prism at total reflection condition Fig. 12.1. A high refractive index prism is coated with a thin metal film (around 50 nm thick) touching the sample. When a light beam is incident through the prism on the surface of thin metal film at total reflection angle θ, the evanescent wave interacts with the sample, and at θ_{sp} angle, it couples with the K_{sp} of surface plasmon. Figure 12.2 shows the dispersion curves of the surface plasmon and the lightwaves. The wave vector K_{ev} of the evanescent wave is equal to the lateral component of the incident lightwave vector (K_g) in the prism.

$$K_{ev} = K_g \sin\theta = \frac{\omega}{c}\sqrt{\varepsilon_g}\sin\theta \tag{12.3}$$

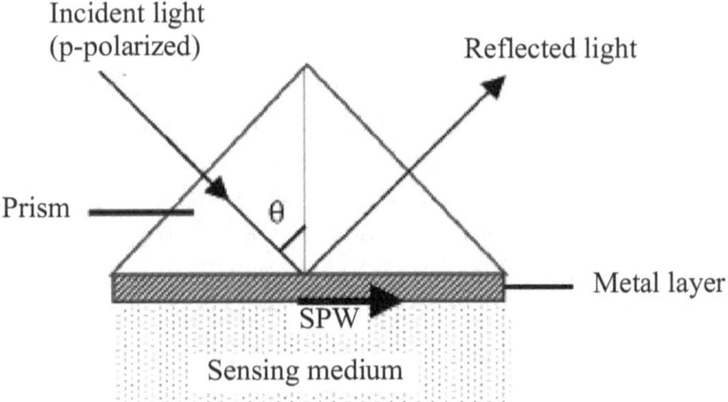

Figure 12.1: Kretschmann configuration of a prism based SPR sensor.

Thus *where* ε_g is the dielectric constant of the prism. The excitation of surface plasmon occurs when the wave vector of the evanescent wave of frequency ω_0 matches that of the surface plasmon of frequency w_0. This occurs at a particular angle of incidence θ_{sp}. This results in the transfer of energy to surface plasmons which reduces the energy of the reflected light. If reflectance R is measured as a function of incident angle θ, a sharp dip is observed at the resonance angle θ_{sp}. A typical R-θ curve is shown in Fig. 12.3. Knowing this angle, the dielectric constant of the sample ε_s can be determined using Eqs. (12.1) and (12.3). The resonance angle is very sensitive to variations in the refractive index (or, dielectric constant) of the sample medium. Moreover, the drop in the resonance is very sharp when the imaginary part of the dielectric constant of the metal is small. The narrower the dip, the better the detection of the variation in refractive index.

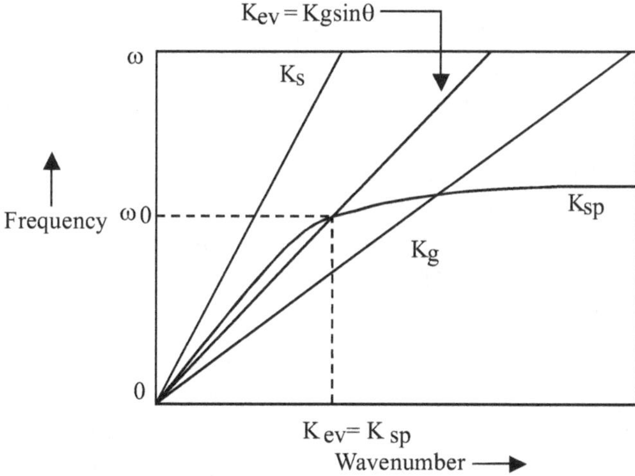

Figure 12.2: Dispersion curves of surface plasmon (K_{sp}), sensing medium (K_s), evanescent wave (K_{ev}) and prism (K_g).

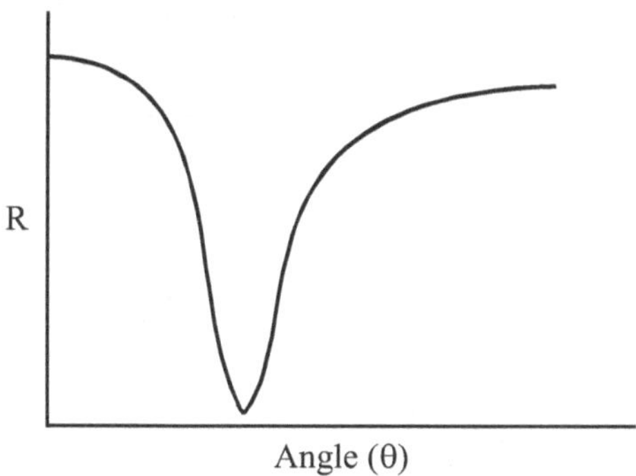

Figure 12.3: Reflectance as a function of angle of incidence at the prism-metal interface.

Both silver and gold have traditionally been used to excite SPR in the visible and near-infrared regions. Silver produces sharper resonance than gold because of the imaginary part of its dielectric constant being less. The permittivity coefficient of the silver and gold layers are –18.5+0.75i and –10.5+2.5i, respectively. Nevertheless, silver has not been used extensively owing to the reactivity of its surface, particularly its susceptibility to oxidation on exposure to laboratory air. The prism based SPR sensing instrument has a number of drawbacks. One of the drawbacks is its bulky size and the presence of various optical and mechanical (moving) parts, which make it difficult to optimize and commercialize on a large scale. Further it is inapplicable for remote sensing applications. Miniaturization of SPR components is a key factor that can be achieved by the use of optical fibers. One of the biggest advantages of using optical fibers is the small diameter of its core which allows them to be used in very small areas. The other advantages of fiber optic SPR sensing configurations are simplified optical design and the capability for remote sensing. Basically, three main approaches have been used in SPR sensors. These are the measurement of SPR-induced intensity change[4], angular interrogation[5] and spectral interrogation[6,7] of SPR.

In the fiber optic SPR sensor the coupling prism is eliminated by depositing the metal film directly on the core of the fiber as shown in Fig. 12.4. The SPR sensing method of fixed angle of incidence and modulated wavelength is selected (wavelength interrogation). This is because the wavelength intensity distribution may be preserved in an optical fiber, whereas the angular intensity distribution of light will be indistinguishable due to mode mixing as a result of the inherent bending of the multimode optical fiber in practical sensing applications. Further, the sensor is fabricated on a multimode fiber there is not a fixed angle of incidence, rather a range of incident angles that are allowed to propagate in the fiber.

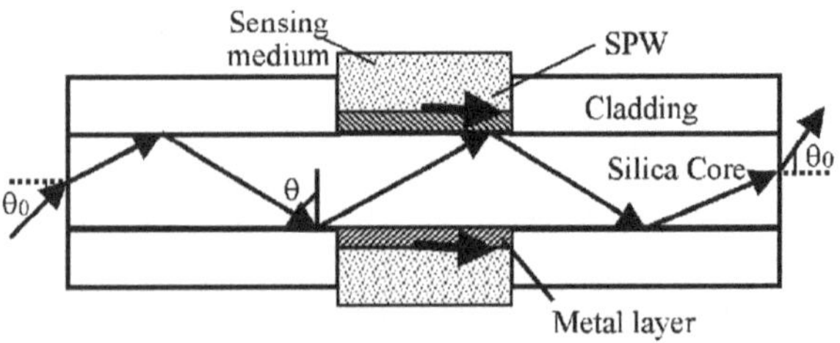

Figure 12.4: Fiber optic SPR sensor probe.

12.3 Modeling of the sensor

The transmitted spectral distribution of intensity of the surface plasmon resonance based fiber optic sensor is affected by the following:

1. Number of reflections of each guided ray in the sensing region
2. The power distribution of guided rays in the fiber.

The number of reflections of guided ray in the sensing region, N_{ref}, is a function of the angle of the ray with the normal to the core-metal layer interface, θ, core diameter, D, and the length of the sensing region, L. The relationship is given by

$$N_{ref} = \frac{L}{D \tan\theta} \tag{12.4}$$

For a given ray, increase in length or decrease in core diameter increases the number of reflections. In order to determine the effective normalized transmitted power at the output end of the fiber, P_{trans}, the reflectance (R) for a single reflection is raised to the power of the number of reflections (N_{ref}) that specific launched angle undergoes with the sensor interface. Therefore, in the case of a guided ray the normalized power transmitted is given by

$$P_{trans} = (R)^{N_{ref}} \tag{12.5}$$

It may be noted that, in the case of an optical fiber, the minimum angle of the guided ray is equal to the critical angle of the fiber. When all the guided rays are launched in the fiber P_{trans} is different for different rays. The width of the spectrum depends on the angle of the ray. In the case of lowest order mode, 90° angle of incidence, the ray travels parallel to the axis of the fiber and does not reflect at the interface ($N_{ref} = 0$) and hence P_{trans} is constant for all wavelengths. In the case of smallest angle of incidence (*i.e.* critical angle) the number of reflections is maximum and the effective spectrum is the broadest[8]. For all guided rays, the SPR spectrum is the sum of the spectrum of all the guided rays. Mathematically the power transmitted is given by

$$P_{trans} = \frac{\int_{\theta_{cf}}^{\pi/2} R^{N_{ref}(\theta)} dP}{\int_{\theta_{cf}}^{\pi/2} dP} \tag{12.6}$$

where dP is the power distribution of the rays,

$$\theta_{cf} = \sin^{-1}\left(\frac{n_{cl}}{n_1}\right) \tag{12.7}$$

is the critical angle of the fiber whereas n_{cl} and n_1 are the refractive indices of the cladding and core respectively. In the multimode optical fiber rays at all possible angles are launched simultaneously using a lens (microscope objective). The numerical aperture (NA) of the lens is chosen larger than that of the fiber so that all the guided rays can be excited in the fiber. The power, dP, arriving at the fiber-end face between the angles θ_0 and $\theta_0+d\theta_0$ is proportional to $(\tan\theta_0/\cos^2\theta_0)\, d\theta_0$, where θ_0 is the angle of the ray from the axis for the ray outside the fiber. Using Snell's law and $\theta = 90^o-\theta_I$, where θ_I is the angle of the same ray from the axis of the fiber but inside the fiber core, dP can be written as[9]

$$dP \propto \frac{n_1^2 \sin\theta\cos\theta}{\left(1-n_1^2\cos^2\theta\right)^2} d\theta \tag{12.8}$$

To calibrate the sensor P_{trans} is determined for different wavelengths and the resonance wavelength, λ_{res} is determined. A plot of λ_{res} as a function of refractive index of the fluid around the metal film gives the calibration curve. The λ_{res} increases as the refractive index increases.

12.4 Sensitivity and signal to noise ratio

As mentioned above, in the SPR sensor based on wavelength interrogation, resonance wavelength (λ_{res}) is determined corresponding to refractive index of the sensing layer (n_s). To obtain λ_{res}, wavelength of the light beam is varied and the corresponding intensity of the reflected light is measured. At λ_{res} sharp dip in reflected light intensity is observed as shown in Fig. 12.5. If the refractive index of the sensing layer is altered by δn_s, the resonance wavelength shifts by $\delta\lambda_{res}$. There are two important characteristic parameters of SPR sensor. These are sensitivity and signal to noise ratio (SNR). Both should be as high as possible for a sensor. The sensitivity (S_n) of a SPR sensor with spectral interrogation is defined as[10]

$$S_n = \frac{\delta\lambda_{res}}{\delta n_s} \tag{12.9}$$

To calculate the sensitivity λ_{res} is plotted as a function of n_s. The slope of this curve gives the sensitivity of the sensor.

If $\delta\lambda_{0.5}$ is the spectral width of the response curve corresponding to 50% reflectivity (Fig. 12.5) and assuming that the detection accuracy of the sensor is inversely proportional to $\delta\lambda_{0.5}$ then SNR of the SPR sensor may be estimated as[11]

$$\mathrm{SNR} = \frac{\delta\lambda_{\mathrm{res}}}{\delta\lambda_{0.5}} \tag{12.10}$$

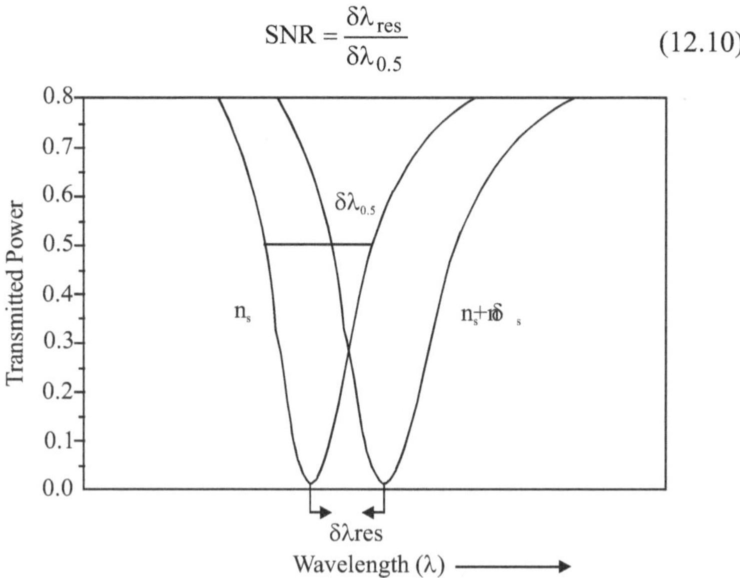

Figure 12.5: Illustration of SPR curve with spectral interrogation. The curves have been plotted for two different refractive indices of the fluid around the metal layers.

As mentioned above, the metallic layer used in SPR measurements consists of either silver or gold. Gold demonstrates a higher shift of resonance parameter to change in refractive index of sensing layer and is chemically stable. Silver, on the other hand, displays a narrower resonance curve causing a higher signal-to-noise ratio (SNR) of SPR chemical sensor, but has a poor chemical stability. The oxidation of silver happens as soon as exposed to air and especially to water, which makes it difficult to get a reliable sensor for practical applications. Treatment of the silver surface by a thin and dense cover is therefore required. Recently, a new structure of resonant metal film based on bimetallic layers (gold as outer one) on the fiber core with angular interrogation sensing has been suggested[12]. The sensitivity and signal-to-noise ratio (SNR) of the optical fiber based bimetallic SPR sensor are evaluated numerically. The effects of fiber parameters like numerical aperture (NA), core diameter and sensing parameters like sensing region length and metallic film thickness have been studied for different ratios of silver and gold layer thickness. The analysis has been extended to remote sensing applications.

12.5 Chemical sensor

The optical excitation of surface plasmon wave at the interface between a thin metal layer and a surrounding medium has been used to develop various chemical and biochemical sensors. The experimental set up of a sensor utilizing surface plasmon resonance excitation in multimode optical fiber is shown in Fig. 12.6. Light from a broadband source is focused onto one of the end faces of the optical fiber having an SPR sensing surface in the middle portion of the fiber. To prepare the sensing element the cladding is removed and a thick silver film of about 550 A° thickness is deposited symmetrically on the fiber core using electron-beam evaporation.

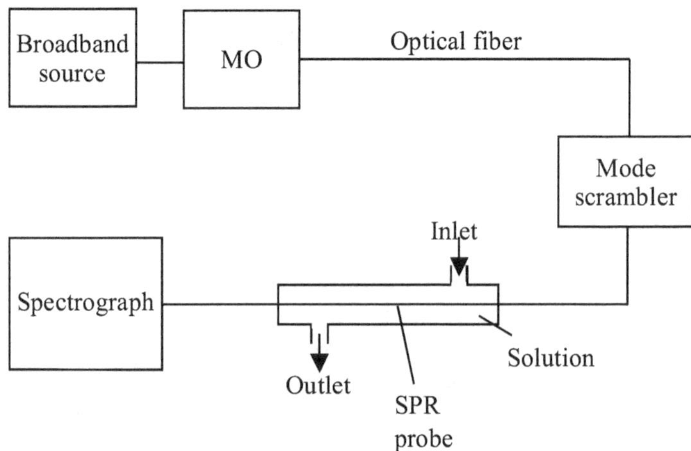

Figure 12.6: A typical SPR based fiber optic sensor.

The SPR sensing surface is enclosed in a flow cell. The output of the fiber is connected to a spectrograph. To launch all the modes, a mode scrambler is used. The solutions of different concentrations and hence refractive indices of a fluid are used to calibrate the sensor. The transmitted spectral intensity distributions are measured for air and all the solutions. The air spectrum is used as a reference to normalize all SPR fiber-optic sensor spectra. Corresponding to each refractive index there exists one resonance wavelength. As the refractive index of the solution increases the SPR coupling or resonance wavelength increases. The variation is non-linear which implies that the sensitivity is wavelength dependent. As the wavelength increases the sensitivity increases. By measuring the wavelength the refractive index of the solution can be determined.

Numerous optical methods have been used in biosensors including fluorescence spectroscopy, interferometry and surface plasmon resonance. In SPR biosensors, antibodies are immobilized on the metallic film. The analyte molecules (antigens) in a solution are captured by antibodies when the sensor is dipped in solution (see Fig. 12.7). This changes the refractive index of the film and hence the intensity of the reflected light from the interface. Therefore, the presence of the analyte can be determined directly without the use of labeled molecules. This makes continuous real-time detection of biomolecular analytes possible.

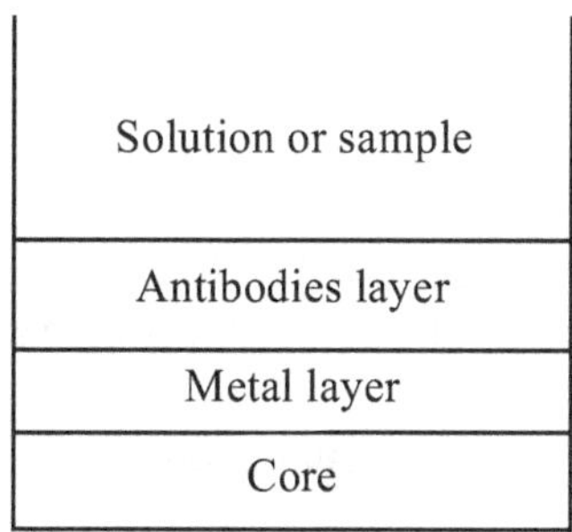

Figure 12.7: Probe configuration for a typical biosensor.

The SPR sensors based on multimode fibers[8,13] exhibit a rather limited resolution mainly due to the modal noise present in multimode fibers causing the strength of the interaction between the fiber-guided light wave and the SPW to fluctuate. To overcome this inherent limitation of SPR sensing devices based on multimode fibers, SPR sensors based on a single mode optical fibers are proposed[14,15]. They include SPR sensors based on tapered[15] and side-polished[14] single mode optical fibers. Out of these two intensity modulated SPR sensors, side polished single mode optical fiber based SPR sensor offers superior sensitivity[15], although suffers from adverse sensitivity to fiber deformations because any deformations in fiber change the state of polarization of the fiber mode and consequently also the strength of its interaction with SPW. To overcome this, a side polished fiber optic SPR sensor based on spectral interrogation using depolarized light is proposed[16]. The applicability of the sensor to refractometry and affinity biosensing is demonstrated. Recently, an optical fiber SPR sensor based on polarization maintaining fibers and wavelength modulation is reported[17]. The design provides superior immunity to deformation of optical fibers of the sensor and hence stability.

Recently, attention has also been focused on introducing optical absorption technique into SPR. An optical absorption-based SPR sensor is proposed and its theoretical aspects are discussed in terms of mathematical descriptions and numerical simulations of the SPR curve[18]. The response theory of the absorption-based SPR sensor is based on the expansion of Kretschmann's SPR theory into the case in which optical absorption in the sensing layer is expressed by the Lorentz model. The theory has now been applied to fibers with single metallic layer[19] and bimetallic layer[20].

References

1 B. Liedberg, C. Nylander and I. Sundstrom (1983) Surface plasmon resonance for gas detection and biosensing. *Sensors and Actuators* B **4**, 299-304.

2 H. Raether, Physics of thin films (Academic : New York, 1984).

3 E. Kretchmann (1971) Die bestimmung optischer konstanten von metallen durch anregung von oberflachenplasmaschwingungen. Z. *Phys.* **241**, 313-324.

4 M.B. Vidal, R. Lopez, S. Aleggret, J. Alonso-Chamarro, I. Garces and J. Mateo (1993) Determination of probable alcohol yield in musts by means of an SPR optical sensor. *Sensors and Actuators B* **11**, 455-459.

5 K. Matsubara, S. Kawata and S. Minami (1988) Optical chemical sensor based on surface plasmon measurement. *Appl. Opt.* **27**, 1160-1163.

6 L.M. Zhang, and D. Uttamchandani (1988) Optical chemical sensing employing surface plasmon resonance. *Electron. Lett.* **23**, 1469-1470.

7 J. Homola, G. Schwotzer, H. Lehmann, R. Willsch, W. Ecke and H. Bartelt (1995) Fiber optic sensor for adsorption studies using surface plasmon resonance. *Proc. SPIE* **2508**, 324-333.

8 R.C. Jorgenson and S.S. Yee (1993) A fiber-optic chemical sensor based on surface plasmon resonance. *Sensors and Actuators B* **12**, 213-220.

9 B. D. Gupta, A. Sharma, and C. D. Singh (1993) Evanescent wave absorption sensors based on uniform and tapered fibers. *Int. J. Optoelectron.* **8**, 409-418.

10 J. Homola (1997) On the sensitivity of surface plasmon resonance sensors with spectral interrogation. *Sensors and Actuators B.* **41**, 207-211.

11 S.A. Zynio, A.V. Samoylov, E.R. Surovtseva, V.M. Mirsky and Y.M. Shirshov (2002) Bimetallic layers increase sensitivity of affinity sensors based on surface plasmon resonance. *Sensors* **2**, 62-70.

12 A.K. Sharma and B.D. Gupta (2005) On the sensitivity and signal to noise ratio of a step-index fiber optic surface plasmon resonance sensor with bimetallic layers. *Opt. Commun.* **245**, 159-169.

13 A. Trouillet, C. Ronot-Trioli, C. Veillas and H. Gagnaire (1996) Chemical sensing by surface plasmon resonance in a multimode optical fibre. *Pure Appl. Opt.* **5**, 227-237.

14 J. Homola and R. Slavik (1996) Fibre-optic sensor based on surface plasmon resonance. *Electron. Lett.* **32**, 480-482.

15 A.J.C. Tubb, F.P. Payne, R.B. Millington and C.R. Lowe (1997) Single-mode optical fiber surface plasma wave chemical sensor. *Sensors and Actuators B* **41**, 71-79.

16 R. Slavik, J. Homola, J. Ctyroky and E. Brynda (2001) Novel spectral fiber optic sensor based on surface plasmon resonance. *Sensors and Actuators B* **74**, 106-111.

17 M. Piliarik, J. Homola, Z. Manikova and J. Ctyroky (2003) Surface plasmon resonance sensor based on a single-mode polarization-maintaining optical fiber. *Sensors and Actuators B* **90**, 236-242.

18 K. Kurihara and K. Suzuki (2002) Theoretical understanding of an absorption–based surface plasmon resonance sensor based on Kretchmann's theory. *Anal. Chem* **74**, 696-701.

19 A.K. Sharma and B.D. Gupta (2004) Absorption-based fiber optic surface plasmon resonance sensor: a theoretical evaluation. *Sensors and Actuators B* **100**, 423-431.

20. B.D. Gupta and A.K. Sharma (2005) Sensitivity evaluation of a multi-layered surface plasmon resonance based fiber optic sensor: a theoretical study. *Sensors and Actuators B* **107**, 40-46.

13

Distributed Sensors

13.1 Introduction

Distributed optical fiber sensors are those sensors which monitor a single measurand at a large number of locations or continuously over the path of the optical fiber. The measurand can be a physical or a chemical quantity. The distributed optical fiber sensor generally consists of a common light source, optical fiber to carry light to the sensing probes and a return fiber to carry modulated light to a photodetector. These sensors can utilize any kind of light modulations described earlier. The advantages associated with distributed sensing are reduced number of components and hence the cost, reduced weight and cabling, and electrical passivity. The applications include the measurements of temperature, pressure, strain, current and chemical concentration. If the measurand is monitored continuously along the fiber length then the sensor is called intrinsic distributed sensor. If it is measured at a finite number of locations then the sensor is called quasi-distributed sensor. The quasi-distributed sensor has a number of advantages over intrinsic distributed sensor. In quasi-distributed sensor, spatial resolution is not as important as in intrinsic distributed sensor provided that individual sensors can be clearly resolved. Further, in quasi-distributed sensing, the sensor elements can be tailored in terms of responsivity and range to meet the specific needs of the application. In this chapter we shall describe various kinds of distributed sensing approaches and few optical fiber distributed sensors.

13.2 Distributed sensing approaches

There are two kinds of approaches namely optical time domain reflectometry and the other is the polarization optical time domain reflectometry. We describe them as :

13.2.1 Optical time domain reflectometry

When light is guided through an optical fiber, transmission loss occurs due to Rayleigh scattering inside the fiber. This is due to the random microscopic variations in refractive index of the fiber core. A fraction of the back-scattered light is returned towards the light source. If an optical pulse is launched in the fiber and the variation of returned back scattered intensity is monitored then the variation in fiber scattering coefficient or attenuation along its length can be determined. This technique is called optical time domain reflectometry (OTDR). This is routinely used in the optical communication industry for the fault detection in the fiber. In sensors, OTDR can be used to detect localized measurand-induced variations in the loss or scattering coefficient of a continuous sensing fiber.

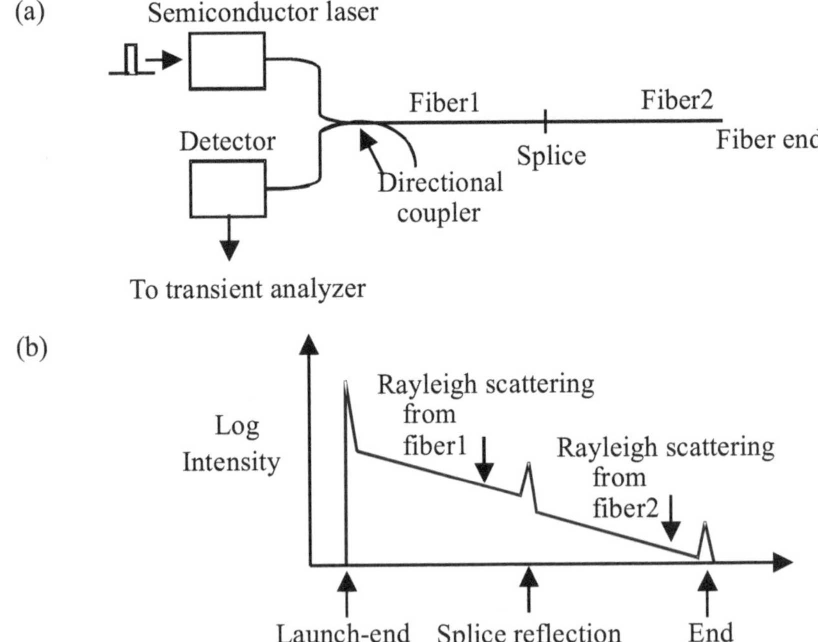

Figure 13.1: Typical OTDR (a) arrangement and (b) its characteristics.

Figure 13.1 shows a typical arrangement for OTDR analysis. A short pulse of light from a laser is launched into the fiber. The intensity of the back-scattered light with time is monitored using a detecting system. If there is no external perturbation, the back-scattered intensity decays exponentially with time due to intrinsic loss. In the presence of external perturbation additional loss of intensity occurs. This is shown in the curve plotted in Fig.13.1. From these measurements the positions of the perturbations and the additional loss at these positions can be found out. The amount of additional loss at each position is the measure of the magnitude of the parameter being monitored at that position. A critical part of the OTDR is the interface between the electronics and the fiber. Signals are transmitted through the fiber at the speed of light, thus the high-speed trans-conductance amplifiers are crucial elements in the sensing portion of the OTDR circuit. Further, OTDR cannot resolve very short distances without complicated and expensive high-speed electronics. Thus some other optical technique is required for high resolution. There are number of sophisticated techniques which can be used such as coherent OTDR, optical frequency-domain reflectometry (OFDR), coherent OFDR *etc*. Each of these techniques improves the conventional OTDR in terms of sensitivity, dynamic range or spectral resolution.

13.2.2 Polarization optical time domain reflectometry

The OTDR technique discussed above is based on the intensity modulation. However, it can also be used with other modulation schemes described earlier. If OTDR relies on the time resolution of the polarization state of Rayleigh back-scattered light in a single mode fiber then the technique is called polarization-optical time domain reflectometry (POTDR). We know that the birefringence parameters of a single mode fiber are sensitive to a number of measurands such as strain, pressure, and electric and magnetic fields. Consequently, the state of polarization of the back-scattered light will vary with distance along the fiber due to these measurands. A narrow pulse from laser is launched into a single mode fiber. The Rayleigh scattering in the fiber does not introduce any change in the mean polarization state of the pulse. Changes in the polarization are introduced by an external measurand field. The POTDR technique appears to be attractive as a way to measure a large number of parameters. However, its main drawback, as with many potentially useful sensing methods, is the variety of parameters to which it may respond. For example, its sensitivity to strain and vibration is troublesome.

13.3 Intrinsic distributed temperature sensor

As defined earlier, in intrinsic distributed sensors the measurand is determined continuously along the fiber path. In this section we shall describe intrinsic distributed sensors for the measurement of temperature. Optical fiber distributed temperature sensors have a wide range of applications in power supply industry, oil and gas industries, process industries, buildings, tunnels and transportations as mentioned below:

A. Power supply industry

1. Fiber optic distributed temperature sensor can monitor the temperature of the power transformer's windings. Thus hot spots, if present, in the windings can be detected and the transformer can be saved from burning.
2. Distributed optical fiber temperature sensor can be used to monitor the temperature of cooling pipes of large power generators. If the cooling pipes are blocked the temperature will increase and it may damage the generator.
3. The capacity of high voltage transmission lines depends on their temperature. If a sensing fiber is installed inside one of the conductors of the transmission lines their temperature can be monitored.
4. In thermal power stations high pressure steam pipes are used. The distributed optical fiber temperature sensor can be used to detect leakage from these pipes.

For all these applications, the fiber must have correct dielectric properties, be nonconductive, non-tracking, non-contaminating and avoid the creation of locally enhanced electric fields. Suitable coatings are required to achieve these objectives and at the same time protect the fiber and withstand the temperature and the chemical attacks they will be subjected to (*e.g.* transformer oil).

B. Petrochemicals (Oil and gas industries)

In petrochemical industries distributed temperature sensor can be used to monitor pipeline temperature. It can also be used to detect gas leakage because gas leakage causes change in temperature.

C. Process industry

The distributed optical fiber temperature sensor can be used to monitor long thermal curing or drying processes in process industry. It can also monitor overheating in machines (*e.g.* in bearings) before damage to the equipment has occurred.

D. Buildings, tunnels, transportation

Fires inside the buildings, tunnels ships and planes occur due to fault in cables. A distributed optical fiber temperature sensor installed in these can detect the outbreak of a fire very quickly wherever it occurs.

E. Construction

Distributed temperature sensors can be used as a temperature measuring device or as a fire alarm in heating and air conditioning units in buildings. They can also monitor the exothermic curing of concrete in dams. This is important because heat weakens the structure. For this the fiber is buried within the structure.

Below we describe some intrinsic distributed temperature sensors based on different phenomena.

13.3.1 Based on Raman scattering

As mentioned in the beginning of this chapter, scattering of the light occurs inside the optical fiber core. The main part of the scattering is Rayleigh scattering that occurs because of changes in the refractive index of the material. In addition to this, the scattered light spectrum also contains a small contribution from Raman and Brillouin scatterings. These scatterings occur due to molecular and bulk vibrations, respectively. The Rayleigh scattering largely does not depend on temperature but the other two types of scattering depend. The wavelengths of the Raman and Brillouin scattered intensities are different from the wavelength of the incident light. The difference in wavelengths reflects a gain or loss of energy by the scattered photon. The shift in Brillouin lines from the incident one is very small. To detect the shift sophisticated detection and source systems are required. In contrast, the Raman spectrum is sufficiently far from the incident wavelength and can readily be separated from Brillouin spectrum using suitable filters. Further, the intensities of these spectral lines depend on the temperature. By selecting one of the Raman lines from the scattered spectrum, the sensitivity of the measurand signal to temperature can be greatly enhanced. In a typical distributed temperature sensor, the shorter wavelength Raman band (the anti-Stokes band) is used to obtain the temperature information. This is because the shorter wavelength Raman band has a higher temperature sensitivity than the longer wavelength Raman band (Stokes band). In the case of germanium-doped silica fiber the anti-Stokes backscattered power increases with the increase in temperature. The detection of Raman backscattered spectrum involves the use of a modified OTDR technique in which the ratio of the Stokes to anti-Stokes intensities of the backscattered light is analysed.

Figure 13.2 shows the experimental arrangement of a distributed fiber optic temperature sensor based on Raman scattering[1]. A short pulse from a high power argon-ion laser (5 W pulses at 514 nm) is launched in the graded index germania doped silica fiber (50 μm/ 125 μm, 0.2 NA) and the ratio of intensities of Stokes to anti-Stokes lines are measured along the fiber path using a photomultiplier tube. The change in the ratio gives the temperature. For Raman OTDR standard multimode telecommunication fiber can be used which is readily available and is relatively cheap. The disadvantage is that the

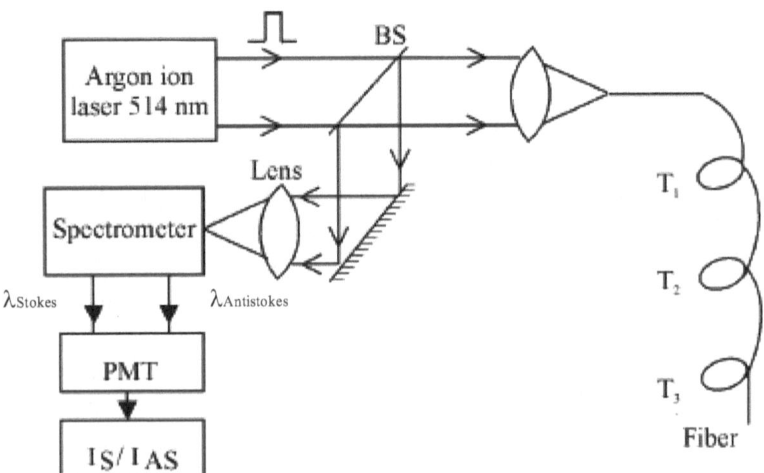

Figure 13.2: Distributed fiber-optic temperature sensor based on Raman scattering. Adapted from ref. 1.

backscattered signal is very weak. This necessitates the use of high input power from the laser and long signal averaging of the detected backscattered signals. The sensor is capable of operation over about 1 km of fiber length with a temperature resolution of about ±1°C and 3 to 10 m spatial resolution.

13.3.2 Based on Rayleigh scattering

The Rayleigh scattering in solid core optical fiber is almost independent of temperature. This is not true in the case of liquid core optical fiber. Since the density and hence the refractive index of the liquid changes with temperature, the Rayleigh scattering will depend on the temperature. This has been exploited to design a distributed temperature sensor[2]. The experimental set up is shown in Fig.13.3. The hollow silica tube filled with ultrapure hexa-chlorobuta 1,3 diene liquid is used as the fiber. The liquid has refractive index higher than that of silica which has no absorption band in the wavelength region of interest. GaAs laser operating at 904 nm is used to launch light in the fiber of numerical aperture 0.3. This fiber couples light into the liquid

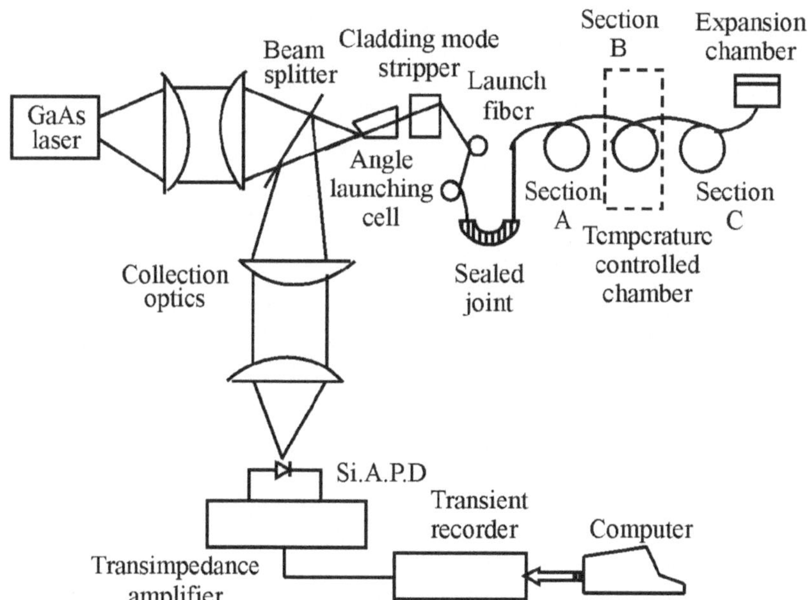

Figure 13.3: Experimental arrangement for the distributed temperature sensor based on Rayleigh scattering[2]. À© [1983] IEEE.

core fiber. The numerical aperture of the liquid core fiber is about 0.5. The increase in temperature of the liquid decreases the numerical aperture of the fiber. The numerical aperture of the liquid core fiber is kept greater than that of the launching fiber for whole range of temperature of interest. The OTDR signals obtained from the liquid core fiber are shown in (Fig.13.4). There are few disadvantages of the sensor. First, the use of liquid core fiber is inconvenient because of the need for expansion reservoir. Second, the dynamic range of the sensor is restricted by the freezing and boiling points of the liquid. Other disadvantages are short term durability, possible void formation under rapid cooling conditions and inability to distinguish between scattering changes due to contamination and temperature variations. In spite of these shortcomings the sensor has ±0.2°C resolution over several hundred meters of the fiber with a spatial resolution of 2m.

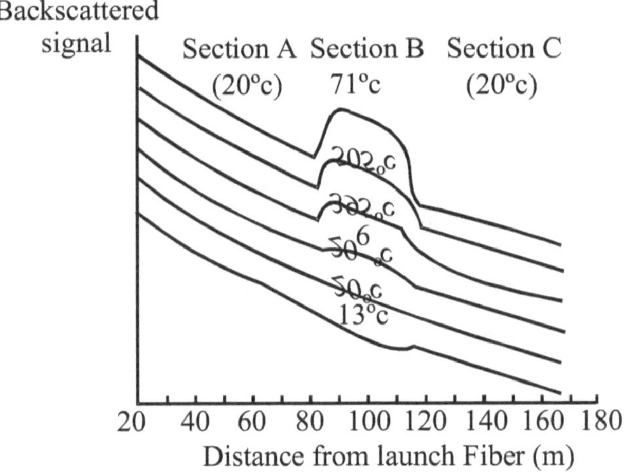

Figure 13.4: Backscattered signal with temperature distribution along the fiber[2]. À© [1983] IEEE.

13.3.3 *Based on temperature-dependent absorption in doped fibers*

Doped multimode optical fibers have also been used to develop distributed temperature sensors. In one of the sensors, the dopant, Nd^{3+}, is added to the core material during the fiber preform fabrication[3]. Nd^{3+} has strong absorption peaks at 740 nm, 800 nm and 880 nm. All the peaks show strong temperature dependence. A linear relationship exists between fiber attenuation and temperature. Further, the percentage change in absorption with temperature increases with dopant concentration. The highest temperature that can be measured depends on the degradation temperature of the fiber coating. For the sensor multimode fiber of 140 m length and 50 μm core diameter with a dopant level of 5 ppm is used. A pulse of 40 ns duration is sent through the fiber from a semiconductor laser emitting radiation at 904 nm and the variation of the backscattered signal with temperature is measured in the range -50 to 100°C. The sensor has a spatial resolution of 15 m and temperature accuracy of 2°C. The sensor can be used in places where high accuracy and resolution are not required (*e.g.* fire alarms in buildings).

13.4 Quasi-distributed sensors

In quasi-distributed sensors, the measurand is not monitored continuously along the fiber, but at a finite number of locations on the fiber path. For example, to make a quasi-distributed temperature sensor, sections of rare-earth doped fibers can be incorporated into a long fiber at certain locations. Or, at certain locations along the fiber the cladding can be modified to respond to the measurand. This technique can measure temperature, refractive index and chemical concentrations. Here we shall describe quasi-distributed sensors for pressure, chemical concentration and temperature.

13.4.1 Pressure sensor based on microbending

Microbending in the fiber is a loss mechanism and therefore it can be used for quasi-distributed sensing of various parameters, such as force, pressure and displacement see Sec. 14.9.1 for microbending sensor). In this case loss can be introduced at any desired location along a fiber by passing the fiber through the jaws of a suitable fiber deformer. The experimental set up employed for the sensor[4] is shown in Fig. 13.5. Light from a laser diode is launched into the fiber which is perturbed by a number of serially placed deformers. In the present case two deformers have been used. A portion of the backscattered light from the 80 ns pulse launched in the fiber is deflected by a beam splitter into an avalanche photodiode. The processing of the output signal is carried out by an oscilloscope interfaced to a computer. The results are plotted using X-Y recorder. The typical results obtained are shown in

Fig. 13.6. For early reception times, the curve has a small slope whose magnitude represents the intrinsic fiber loss. This slope is constant throughout the fiber except at the locations of two deformers where the jumps indicate the loss due to pressure applied. Converting time scale into fiber length one can find the locations and the pressure applied at these two locations.

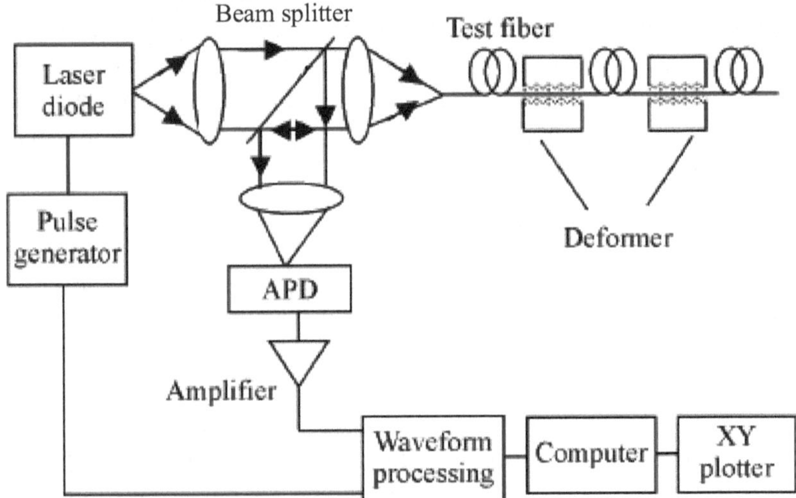

Figure 13.5: Experimental arrangement for quasi-distributed pressure sensor based on microbending[4]. À© [1984] IEEE

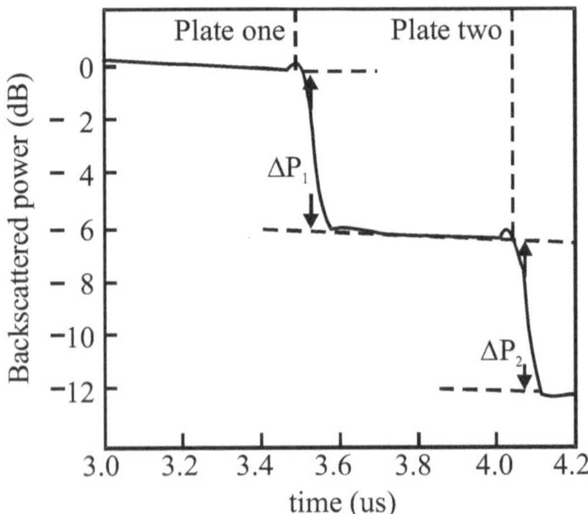

Figure 13.6: Backscattered signal as a function of time[4]. À© [1984] IEEE.

13.4.2 Chemical sensor

Distributed chemical sensors can be fabricated by modifying the fiber cladding so that its light transmission properties are altered through its contact with the particular species of interest. Sensing mechanisms that have been utilized include evanescent wave absorption, Rayleigh scattering, Raman scattering, fluorescence and fluorescence quenching. Figure 13.7 shows the experimental set up of a quasi-distributed optical fiber chemical sensor[5]. OTDR technique and the evanescent wave absorption mechanism are used for the sensor. A 25 m long PCS fiber is used to fabricate the sensor. At each sensing region a 25 cm of cladding is removed and the pH sensitive dye (cresol red) is immobilized on the core at these locations. Sensing regions are positioned at 5m, 10m and 20m from the end of the fiber which is then butt jointed onto the rest of the fiber optic link. A 5ns pulse from the nitrogen pumped dye laser is launched into the fiber through a beam splitter. A fast photodiode is used to provide a trigger signal for the data acquisition system. To launch the light into the fiber and to collect the backscattered light an aspheric lens is used. The unwanted fluorescence signal in the backscattered light is filtered using a monochromator. The backscattered light is detected by a photomultiplier tube which is then processed by a transient digitizer. The absorption properties of dye used change with the change in the pH of the surrounding. Thus the three sensing regions prepared can change state when exposed to acid or alkaline vapour and hence the backscattered signal will get affected. From the measured signal one can find out the pH of the surroundings at three locations.

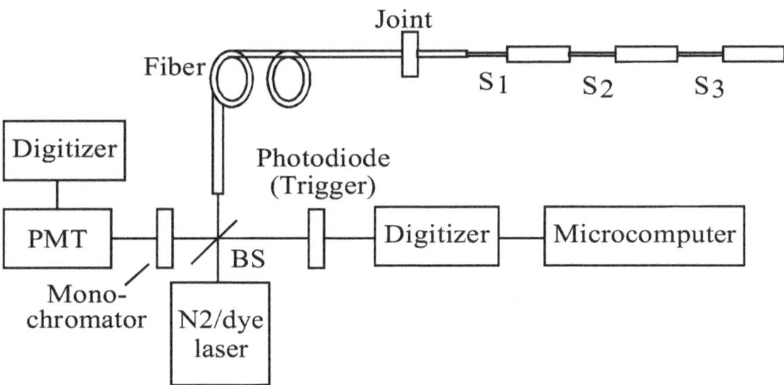

Figure 13.7: Quasi-distributed chemical sensor based on evanescent wave absorption[5].

Another distributed fiber optic chemical sensor utilizing evanescent wave is reported recently for hydrogen gas[6]. The silica core of the fiber is coated with platinum supported tungsten trioxide thin film as hydrogen sensitive cladding. The optical power propagating in the core region leaks to the cladding and is susceptible to color change due to the reduction of tungsten trioxide by hydrogen. The propagation loss caused by the exposure of the sensor to hydrogen gas is found to be remarkable and the sensitivity is strongly dependent on the sensor length. The ability of simultaneous detection of hydrogen with location identification is determined using OTDR technique after splicing three fiber optic sensors in series along the fiber cable.

13.4.3 Strain/Temperature sensor based on Bragg grating

In chapter 11, fiber Bragg grating based sensors have been described to sense strain and temperature. To measure these at different locations a number of Bragg gratings at these locations are written onto the core of the Ge-doped fiber Fig. 13.8. If a pulse from a broadband source is launched into the fiber then each grating will back-reflect light whose wavelength will depend on the grating spacing. The grating spacing depends on the temperature and strain. By measuring the back-reflected wavelengths with time the temperature or the strain can be measured at different locations. The distributed strain sensor has applications in monitoring structures such as plateforms, pipelines, ships *etc*. Strain and crack growth in bridges, dams, tall building *etc*. can also be monitored by distributed strain sensor.

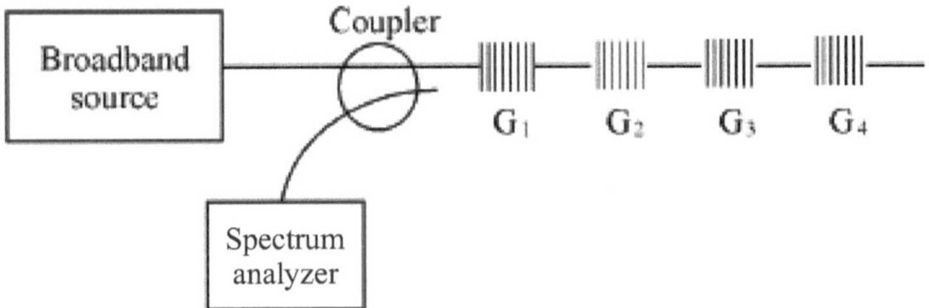

Figure 13.8: Quasi-distributed temperature/strain sensor. Gs are the FBGs.

13.4.4 Water sensor

The presence of water at different locations can also be detected using a single fiber. The technique is the same as used above. The chemically sensitive water swellable polymer, hydrogel, is used as the active sensing component. Hydrogels are a range of hydrophillic macromolecular polymers that absorb water and swell in aqueous media without dissolution. The swelling action is converted into a mechanical response in the form of a force or pressure. Mechanical perturbations or bends in the fiber can cause guided modes within the fiber core to be coupled to the cladding where they are severely attenuated. These regions of loss can be located by the use of OTDR technique. The water sensor based on this principle[7] is shown in Fig. 13.9. The hydrogel material is deposited onto a central supporting former (rod). It is then held in contact with an optical fiber by a helically wound Kevlar wrap with a 2 mm winding pitch. Acrylate-coated graded index multimode fiber is used to fabricate the sensor. In the presence of water the hydrogel swells and exerts a microbending force through the Kevlar wrap onto the fiber. The microbending effect causes a loss of power in the vicinity of the water and thus enables the water and its location to be detected using OTDR. The sensor is capable of detecting water ingress points as small as 50 cm in length over lengths of sensor in excess of 100 m.

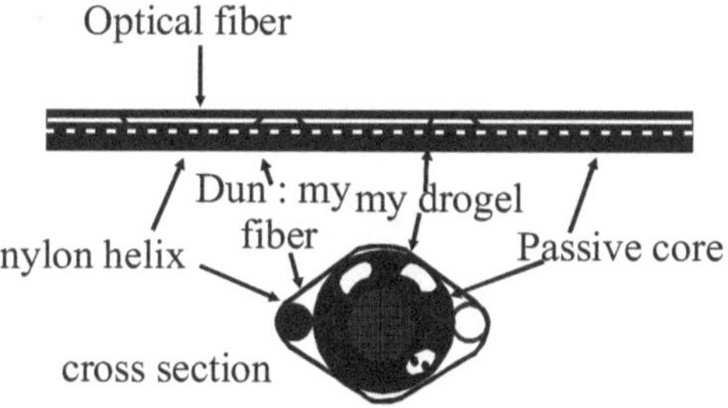

Figure 13.10: Schematic of hydrogel based water sensor[7]. À© [1995] IEEE.

References

1. J.P. Dakin, D.J. Pratt, G.W. Bibby and J.N. Ross (1985) Distributed anti-Stokes ratio thermometry. 3rd Int. Conf. on Optical Fibre Sensors, San Diego (USA).

2. A.H. Hartog (1983) A distributed temperature sensor based on liquid core optical fibers. *IEEE J. Lightwave Technol.* **LT-1**, 498-509.

3. M.C. Farries, M.E. Fermann, R.I. Laming, S.B. Poole, D.N. Payne and A.P. Leach (1986) Distributed temperature sensor using Nd^{3+}- doped optical fibre. *Electron. Lett.* **22**, 418-419.

4. A.R. Mickelson, O. Klevhus and M. Erikssud (1984) Backscatter readout from serial microbending sensors. *IEEE J. Lightwave Technol.* **LT-2**, 700-709.

5. F. Kvasnik and A.D. McGrath (1989) Distributed chemical sensing utilising evanescent wave interactions. *Proc SPIE* **1172**, 75-82.

6. S. Sumida, S. Okazaki, S. Asakura, H. Nakagawa, H. Murayama and T. Hasegawa (2005) Distributed hydrogen determination with fiber optic sensor. *Sensors and Actuators B.* **108**, 508-514.

7. W.C. Michie, B. Culshaw, M. Konstantaki, I. McKenzie, S. Kelly, N.B. Graham and C. Moran (1995) Distributed pH and water detection using fiber-optic sensors and hydrogels. *IEEE J. Lightwave Technol.* **LT-13**, 1415-1420.

14

Miscellaneous Sensors

14.1 Introduction

In this chapter we shall describe various simple and easy to fabricate fiber optic sensors. Most of these sensors are based on intensity modulation. Intensity modulated optical fiber sensors have been extensively studied in comparison to sensors based on other modulation techniques. This is because of their several advantages such as simplicity, reliability, flexibility and relatively low cost. Intensity modulated sensors make use of multimode fibers. Multimode fibers have large core and high numerical aperture that make the coupling of light into the fiber easier. The large core is also better suited to propagate high power laser beams. Further, the plastic-clad fused silica multimode fibers are available. Plastic-clad silica multimode fibers have an additional advantage. Their plastic cladding can be removed easily permitting access to the core. The evanescent field in the unclad part has been used for various kinds of chemical and biochemical sensors (see chapter 5). Above all, these fibers are cheap. This chapter is devoted to simple and easy to make both intrinsic and extrinsic type sensors. Sensors utilizing bare-ended fibers have also been described.

14.2 Displacement sensor

Fiber optic displacement sensors are useful for industrial, military and medical applications. The main advantages of fiber optic displacement sensors over the conventional ones are high accuracy, possible use in composite structures and in some cases of non-contact sensing. Transmissive, reflective and microbending concepts have been used to measure displacement. In transmissive concept light is directly coupled from one fiber to another while in reflective concept it is coupled after reflection from the surface of which the displacement is to be measured. In microbending concept, the power transmitted is lost when the fiber is bent. The specific applications of the displacement sensors are in the measurement of axial motion, proximity, eccentricity, vibration and shaft run out. These sensors can also be used for the measurement of film thickness, rotation and parts gauging. In this section transmissive and reflective concepts based fiber optic displacement sensors will be described. Sensor based on microbending loss is described in Sec. 14.9.1.

14.2.1 Transmissive concept

Figure 14.1 shows the basic configuration of a fiber optic axial displacement sensor based on transmissive concept. There are two fibers having same axis, one is called transmitting and the other is receiving fiber. Light is launched into the transmitting fiber. The receiving fiber collects a fraction of light exiting from the other end of the transmitting fiber. The light guided by the receiving fiber is detected at the other end. As the distance between the two fibers (R) increases the amount of light collected by the receiving fiber decreases. A typical variation of normalized intensity as a function of displacement (or the separation) of the fiber is shown in Fig.14.2.

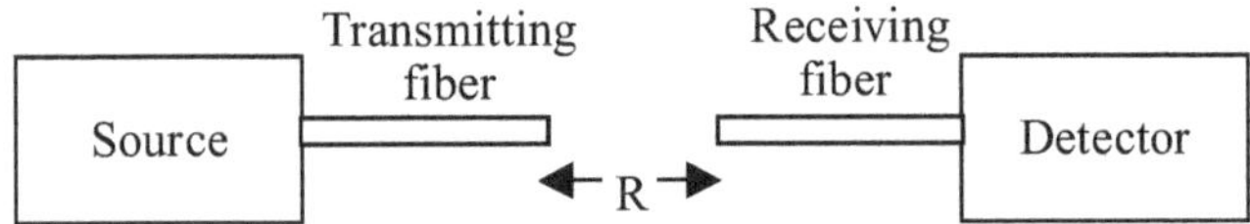

Figure 14.1: Fiber optic displacement sensor based on transmissive concept.

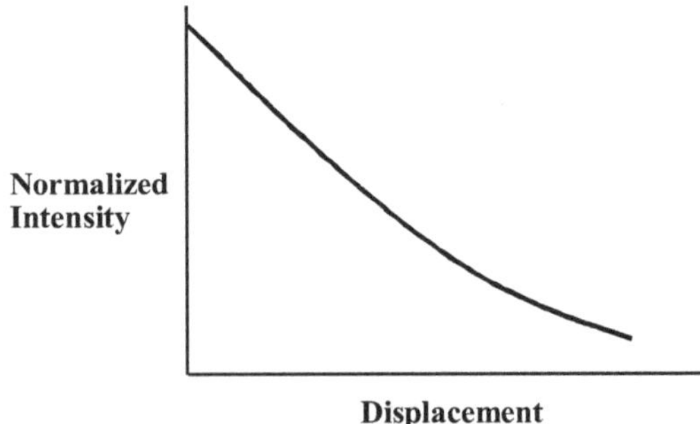

Figure 14.2: A typical normalized transmitted intensity versus displacement curve of the sensor based on transmissive concept.

The variation in intensity with displacement can be explained on the basis of geometrical optics. To explain, let us consider two identical fibers having same axis (see Fig. 14.3). Light is launched into fiber 1. Light exits from fiber 1 in the form of a solid cone defined by its numerical aperture. Let the receiving fiber 2 is at a distance R from the tip of the fiber 1. If 2ρ is the diameter of the fiber1 and R>>2ρ then

$$\tan\theta = \frac{D}{2R} \tag{14.1}$$

where θ and D correspond to a spot in the plane of the tip of the fiber 2 as shown in Fig. 14.3. The angle θ is related to the numerical aperture of the fiber. The area of the spot is given by

$$A = \pi\left(\frac{D}{2}\right)^2 = \pi R^2 \tan^2\theta \tag{14.2}$$

If P is the power of the light emitted from fiber 1 then the intensity at a distance R is given by

$$I(R) = \frac{P}{A} = \frac{P}{\pi R^2 \tan^2\theta} \tag{14.3}$$

From Eq. (14.3) it can be seen that for a given fiber and the light source the intensity of the light collected by the receiving fiber is inversely proportional to the square of its distance (or displacement) from fiber 1. The variation is similar to that shown in Fig. 14.2. To obtain the best performance of the sensor, fiber ends should be flat and polished. Further, the axes of both the fibers should be the same.

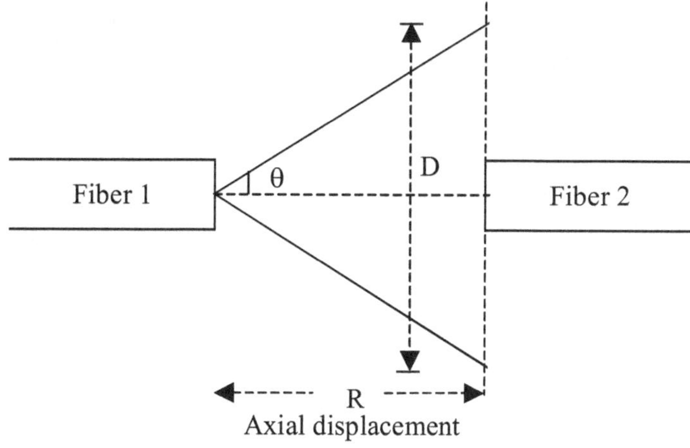

Figure 14.3: Geometry used to relate the axial displacement 'R' to the intensity.

The arrangement mentioned above can also be used to measure the radial displacement *i.e.* the displacement along the direction perpendicular to the fiber axis (see Fig. 14.4). In this case the intensity collected by the fiber 2 decreases as the fiber moves away from the axis of the fiber 1.

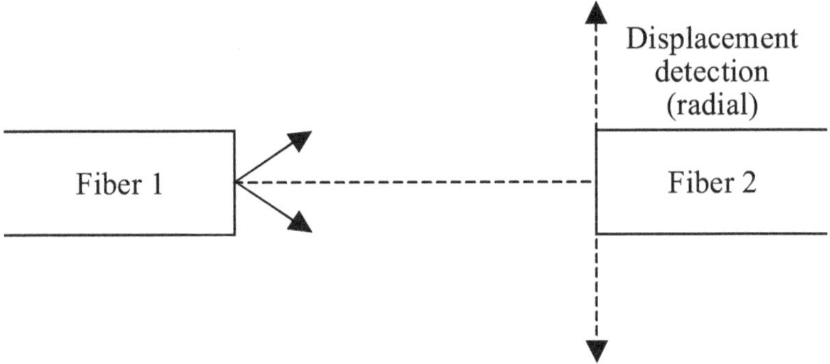

Figure 14.4: Radial displacement sensor based on transmissive concept.

14.2.2 *Reflective concept*

The displacement sensors can also be designed on the basis of reflective concept. Figure 14.5 shows the basic configuration of a fiber optic displacement sensor based on the movement of a mirror external to the fiber. Light is launched into the transmitting fiber and is incident upon the mirror. The receiving fiber then collects a fraction of the reflected light and is transmitted to the detector. As the distance between the mirror and the fiber pair increases the amount of light collected by the receiving fiber changes. Thus by measuring the power collected by the receiving fiber, the displacement of the mirror can be found out.

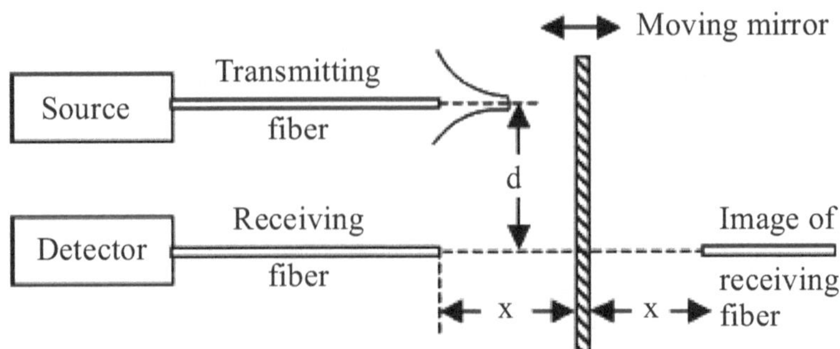

Figure 14.5: Fiber optic displacement sensor based on reflective concept.

To describe the output at the detector we consider an image of receiving fiber in the mirror (Fig. 14.5). The problem is now to find the propagation of light from transmitting fiber over the distance 2x and then how much light will be collected by the image of the receiving fiber; x is the separation between transmitting fiber and the mirror. When light is launched into the transmitting fiber all modes are excited at the input end. As light propagates down the fiber higher order modes (larger angles) are more highly attenuated than the lower order modes (smaller angles). This attenuation reduces the total amount of power propagating through the fiber as well as changes the distribution of power between modes. The stable or equilibrium position comes when power decreases as the mode angle increases. Once this is achieved, the intensity profile at the output end of the transmitting fiber is close to a Gaussian function. The beam then propagates toward the mirror that reflects and sends it back to receiving fiber. Mathematically, the Gaussian beam profile at a distance z from the exit point of the transmitting fiber can be written as[1]

$$I(d,z)=I_o \exp\left[-\frac{2d^2}{{d_o}^2}\left(1+\frac{z^2}{{z_R}^2}\right)^{-1}\right]\cdot\left(1+\frac{z^2}{{z_R}^2}\right)^{-1} \tag{14.4}$$

where I_o is the irradiance at the centre of the Gaussian beam waist; d_o is the beam waist diameter; d is the radial distance from the beam centre or the spacing between the transmitting and receiving fibers and z_R is the Rayleigh range. If d is defined as the beam diameter at e^{-2} irradiance point then

$$d^2 = d_0^2 + \theta^2 z^2,$$
$$z_R = d_o / \theta,$$
$$\sin\theta = NA$$
$$z = 2x$$

Thus the amount of light received by the detector (I) depends on d and the position of the mirror (x). A typical normalized intensity versus separation between transmitting fiber and the mirror is shown in Fig. 14.6. It can be seen from the figure that for two different positions of the mirror the normalized intensity is the same. Thus for this design either the mirror should remain on the left side of the peak or to the right side of the peak. Which side should be used will depend on the desired operating range of the sensor.

The response curve of the sensor can also be explained on the basis of geometrical optics (Fig. 14.7). Consider a ray emerging from the fiber 1. If the mirror is very close to the tip of the fiber1 (position 1) then the ray will not couple into the fiber 2. As the separation increases the coupling starts (position 2) and the maximum coupling occurs when most of the rays are coupled into the fiber 2. With further increase in the separation intensity received decreases. The position of the peak depends on the separation between the fibers.

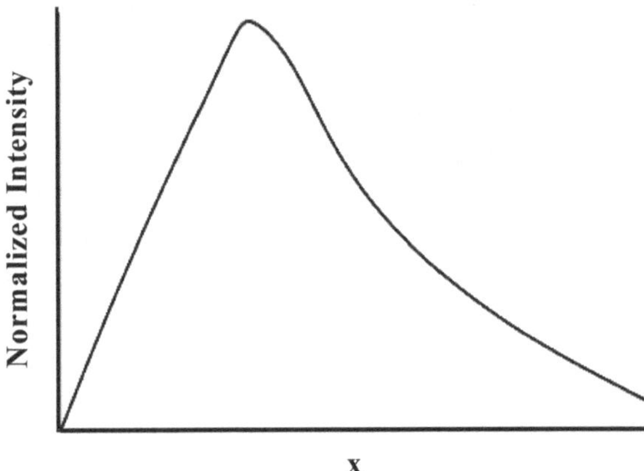

Figure 14.6: A typical reflection sensor response curve.

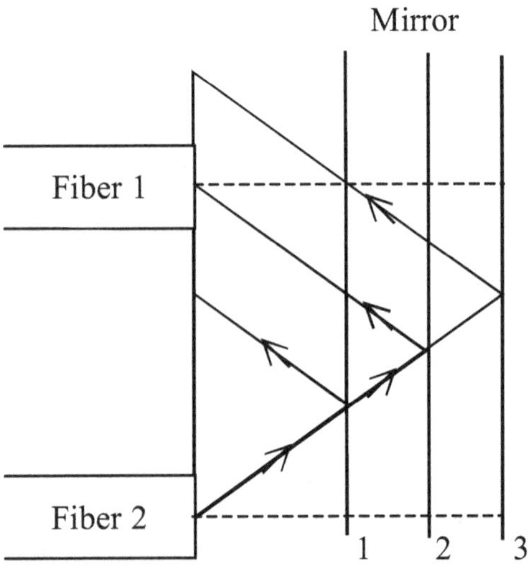

Figure 14.7: Geometrical optics explanation of the response curve of the displacement sensor utilizing reflective concept.

The double valued response curve of the sensor can be removed if a Y-type of configuration is used as shown in Fig. 14.8. In this case response obtained is similar to the sensor based on transmissive concept.

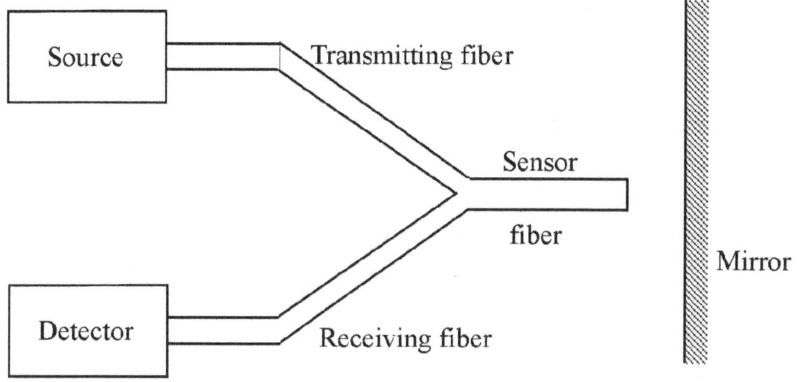

Figure 14.8: Single optical fiber displacement sensor configuration based on reflective concept.

14.3 Flow measurement

Optical fibers can be used to measure the liquid flow in a tube. The principle is based on the interruption of a light beam due to the presence of an object.

The sensor consists of two fibers directed at each other (Fig. 14.9). One transmits light while the other receives it. The fiber transmitting

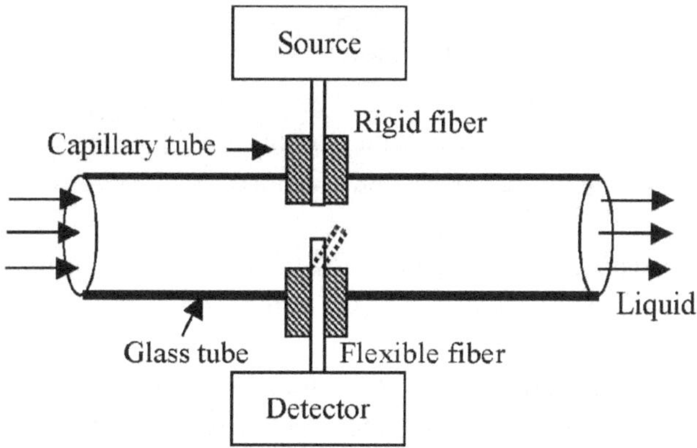

Figure 14.9: Liquid flow sensor.

light is rigid and the one receiving it is flexible. If a liquid is flowing through the tube then the flexible fiber will bend and the amount of light received by it will decrease. The decrease in intensity will depend on the flow velocity of the liquid. The sensor is based on the transmissive concept described above.

14.4 Acoustic sensor

Figure 14.10 shows a fiber-optic acoustic sensor. The principle is the same as that of liquid flow sensor. It consists of two multimode fibers, one is called input while the other is output fiber. Light is launched into input fiber whose other end is kept rigid. The output fiber having same axis as the input receives the light exiting from the rigid end of the input fiber. The end of the output fiber that collects light is kept in contact with a diaphragm of the speaker. Its other end is connected to a detecting system. The motion of the diaphragm displaces the end of the output fiber with respect to the beam of the light exiting from the input fiber. This causes the light collected by the output fiber and hence the detector to vary according to the movement of the diaphragm.

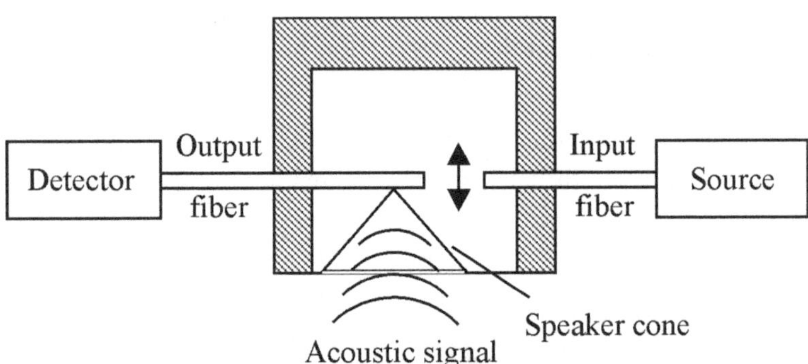

Figure 14.10: Acoustic sensor based on transmissive concept.

The acoustic sensor is well suited to air in which normal acoustic intensities produce relatively large vibrational amplitude of molecules. For a microphone working in water i.e., a hydrophone, the fluid density is high. This will produce smaller vibratory motion for the same acoustic power level. Thus, in water more sensitivity is required. The hydrophones based on phase modulation are highly sensitive. These can detect acoustic noise at great depth. An acoustic sensor based on phase modulation is described in Section 8.6.

14.5 Refractive index sensor

The refractive index of a liquid can be measured using optical fibers and intensity modulation. There are number of ways of measuring it. One of the sensors in the category of external sensors is shown in Fig. 14.11. Light is launched into fiber 1 and is coupled into fiber2 which has the same axis as fiber 1. The light is then detected by the detector connected at the other end of the fiber 2. For a fixed gap between two fibers, the light collected by the fiber 2 depends on the numerical aperture of the output end of the fiber 1 and hence on the angle of the cone of the light emitted from it. If the refractive index of the medium between two fibers increases, the numerical aperture will decrease and the angle of the cone (θ) will also decrease. This will result in the increase in the light coupled into the fiber 2. The ratio of the power collected by fiber 2 in the presence of a specimen liquid to that in the presence of air is a linear function of the refractive index of the liquid. The configuration has applications in the detection of acid level in batteries and a threshold liquid level.

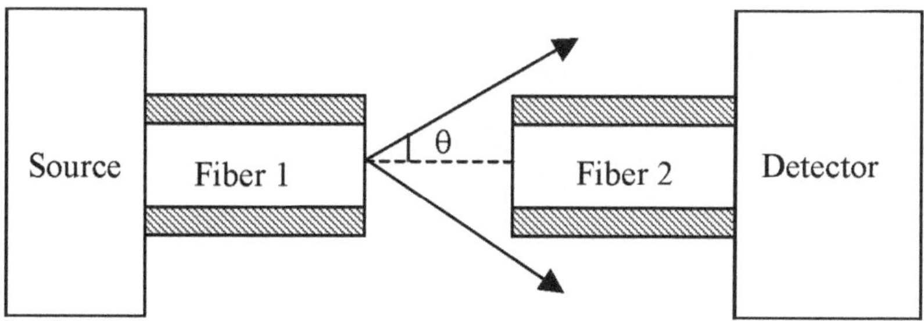

Figure 14.11: A fiber optic refractive index sensor based on transmissive concept.

The refractive index sensor described above is external type sensor. There are a number of intrinsic type refractive index sensors. These are based on different kinds of probes like straight and uniform, tapered and U-shaped. Below we describe them in detail.

14.5.1 Uniform and straight unclad probe

This is the simplest intrinsic sensor for the measurement of the refractive index of the liquid. It is based on the principle of light guidance in the fiber. It is similar to the fiber optic evanescent field absorption sensor described in Section 5.6. The only difference is that in this case the unclad portion is immersed in liquids of different refractive indices. These liquids are non-absorbing. The measurement procedure is the same. The normalized output power as a function of refractive index of the liquid for a PCS fiber of 600 μm core diameter and 0.17 numerical aperture is shown in (Fig. 14.12). It may be seen from the figure that as long as the refractive index of the liquid (n_{liq}) is less than the refractive index of the cladding (n_{cl}) no appreciable change takes place in the power output. This is because, for $n_{liq}<n_{cl}$, the unclad portion's numerical aperture is greater than that of the fiber itself. That means it is highly or strongly guiding. All the rays in this region suffer total internal reflections. For $n_{liq}>n_{cl}$, the critical angle with respect to the normal to the core-cladding interface of the unclad region ($\sin^{-1}(n_{liq}/n_{co})$) is greater than that of the fiber ($\sin^{-1}(n_{cl}/n_{co})$) and therefore all the rays do not suffer total internal reflection in this region: n_{co} is the refractive index of the fiber core. Some of the rays get refracted and hence the output power decreases. The decrease in power increases with the increase in the refractive index of the liquid. For $n_{liq} = n_{co}$, the output power detected

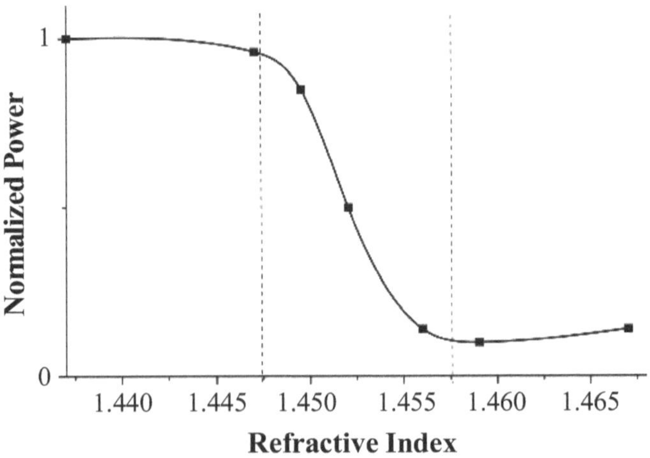

Figure 14.12: A calibration curve of a fiber optic refractive index sensor using unclad fiber as the probe.

is almost zero. The sensor can be used to measure the refractive index of the liquid from n_{cl} to n_{co}. In the present case the range is very small. To increase the range fibers of high numerical apertures should be used.

14.5.2 Tapered probe

The refractive index range of the sensor can be increased if the unclad region of the fiber is tapered. In the previous sensor, for $n_{liq} < n_{cl}$, the sensing region was highly guiding. In the case of tapered probe it is not true. This is because as the ray enters in the tapered region its angle with the normal to the interface decreases as it propagates see Fig. 5.6. If the angle becomes less than the critical angle of the taper region then the ray will get refracted. Thus the ray that was guided in the straight and uniform core sensor may not be guided in tapered sensor. Therefore, even at low refractive index, one can see the change in the output power. If P_o is the total power injected into the fiber then the power transmitted is given by[2]

$$P = P_o \frac{n_{co}^2 - n_{liq}^2}{R_T^2 \left(n_{co}^2 - n_{cl}^2\right)} \tag{14.5}$$

where R_T ($= \rho_i/\rho_o$) is the taper ratio. According to this equation the power output increases linearly with the decrease in n_{liq}^2. Further, the range of the sensor increases with the increase in the taper ratio, R_T.

14.5.3 U-shaped probe

The range of the fiber optic refractive index sensor can also be increased by bending the unclad portion of the plastic-clad silica fiber into the form of an U-shape see Fig. 5.8. The bent section of the fiber, when immersed in a liquid of refractive index n_{liq} ($> n_{air}$) yields a signal different from that in a bare fiber. This is because, as the ray enters into the bent region, its angle with the normal to the interface decreases. The decrement depends on the bending radius of the U-shape. Smaller the radius, larger is the decrease in angle. Hence the ray that was guided in the straight region of the fiber may not remain guided in the bent region. The sensitivity and the dynamic range of the sensor depend on the bending radius.

14.6 Detection of oil in water

The principle of detection of oil in water is based on the light guiding property of the fiber. That is, the increase in the refractive index of the cladding decreases the power transmitted. The experimental set-up is shown in Fig. 14.13.

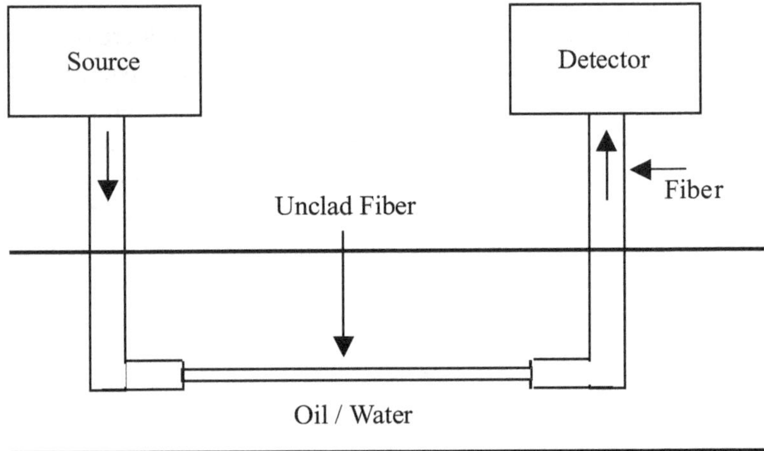

Figure 14.13: Experimental set-up to detect oil in water.

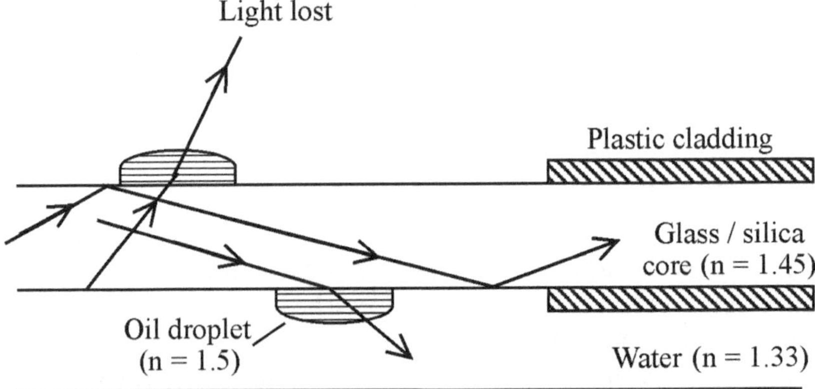

Figure 14.14: The principle of sensor to detect oil in water.

A small length of the cladding from the middle portion of the fiber is removed. The unclad part of the fiber is immersed into a container filled with water. Light is launched into one of the ends of the fiber and all the modes are excited. One can use a He-Ne laser with a microscope objective having NA higher than that of the fiber to launch the light and to excite all the modes. The light exiting from the other end of the fiber is measured by means of a power meter. If the oil is added in the container it gets attached to the core. Since the refractive index of the oil is more than that of the water and in some cases greater than the core, the loss in transmitted power occurs Fig. 14.14. Thus one can get the information of oil leakage. The disadvantage of the sensor is that the waxy oil gets attached to the surface of the fiber core and one has to use some detergent every time to remove it. The same principle can be used to detect the kerosene content in petrol and also the concentration of sugar solutions. In all these cases a change in refractive index occurs.

14.7 Liquid level sensor

The detection of liquid level is one of the many functions for which fiber optic sensors are often more suitable than their electrical counterparts, especially in noisy or hostile environments. Commercially, most attention has been focused on a simple on/off liquid level switch. One of the liquid level sensors developed is based on the total internal reflection phenomenon. It involves two fibers and a 90° glass microprism. The fibers are connected to the base of the prism as shown in Fig. 14.15. The light is launched into one of the fibers. When the

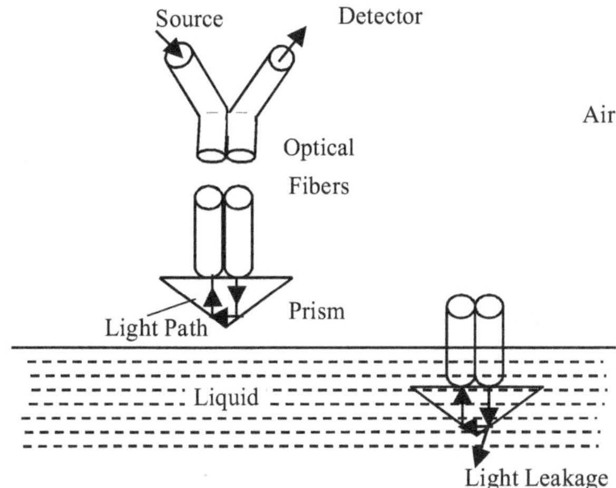

Figure 14.15: Liquid level sensor based on total internal reflection.

prism is surrounded by air, the light faces total internal reflection at the sides of the prism. The light then coupled to the second fiber which is connected to a detector. When the prism comes in contact of the liquid, light suffers attenuation due to frustrated total internal reflection. This results in a drop in the signal level at the detector.

One can also discriminate between different liquids such as gasoline and water by using proper electronic circuit. This is because the amount of light lost from the probe is a function of the refractive index of the liquid.

14.8 Temperature sensors

In this section we shall describe both intrinsic and extrinsic type fiber optic temperature sensors based on different phenomenon.

14.8.1 Temperature dependent refractive index of the liquid

The refractive index of most of the oils decreases with the increase in their temperature. Based on this, a temperature sensor has been developed [3]. The cladding from one of the ends of a multimode optical fiber is removed. The unclad portion of the fiber is inserted in a capillary tube containing oil (Fig.14.16) and a mirror is fixed at the end. The mirror reflects the light exiting from the end of the fiber. The light is then recoupled to the same fiber. As the temperature of the environment

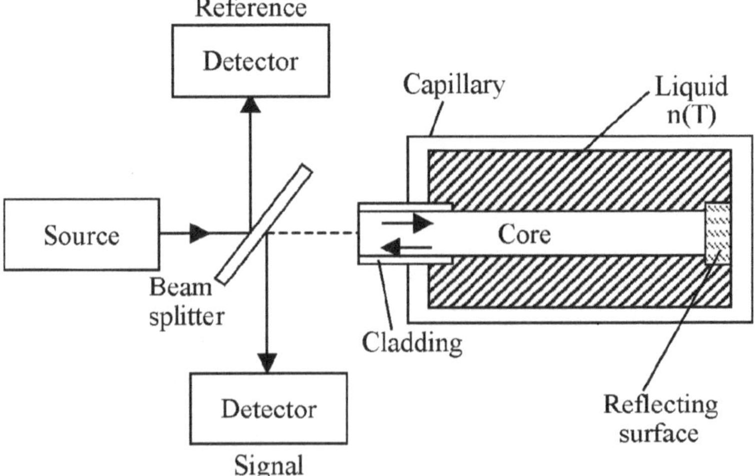

Figure 14.16: Fiber optic temperature probe based on change in refractive index.

changes the refractive index of the oil changes thereby leading to an effective change in the numerical aperture of the fiber. This change affects the intensity of the light received by the detector after reflection from the mirror. Thus the temperature of the outside can be determined by comparing this intensity with the intensity reflected when the probe is in contact with a reference temperature source. The typical response curve of the sensor is shown in Fig. 14.17. By choosing appropriate liquids or a mixture of liquids the operating temperature range of the sensor can be altered.

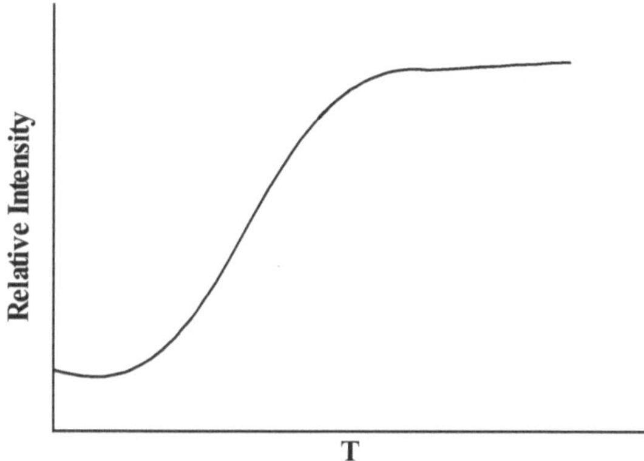

Figure 14.17: A typical response curve of the sensor based on refractive index change.

14.8.2 Thermal expansion of core and cladding

At room temperature, the standard dual coated optical fiber used for communication networks has a soft primary coating and a stiff secondary coating. This combination prevents physical damage to the fiber. If the temperature of the fiber is decreased the primary coating also becomes stiffer. The decrease in temperature results in the decrease in the specific volume of the materials. This induces the microbending of the fiber and hence the attenuation of light propagating within the fiber. The experimental set up of a cryogenic temperature sensor based on this[4] is shown in Fig. 14.18. Dispersion-shifted, single mode optical fiber with a special acrylate dual coating is used in the experiment. Light from a monochromatic source is launched into a fiber coupler. One end of the coupler is connected to a detector while the other end passes through the cryostat and then connected to another detector. The fiber is cooled to 20K and then heated to room temperature while measurements are made. The microbending loss with temperature is shown in Fig.14.19. It can be seen that as the temperature increases the loss decreases. For temperature greater than 200K no loss is observed. This is because the glass transition temperature of the primary coating is around 200K. The sensitivity of the sensor increases with the increase in the fiber length or the increase in the wavelength of the source. The dynamic range of the sensor is from 20K to 180K. For accurate measurements the wavelength of the source should remain stable.

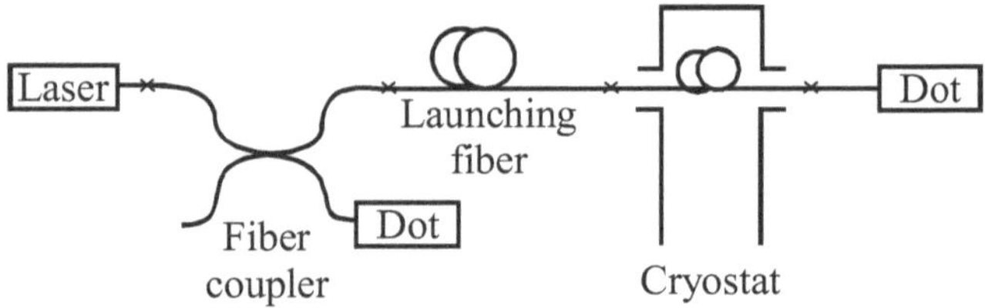

Figure 14.18: Experimental set up of a cryogenic temperature sensor. The crosses symbolize splice points. Reprinted from ref. 4 with permission from OSA.

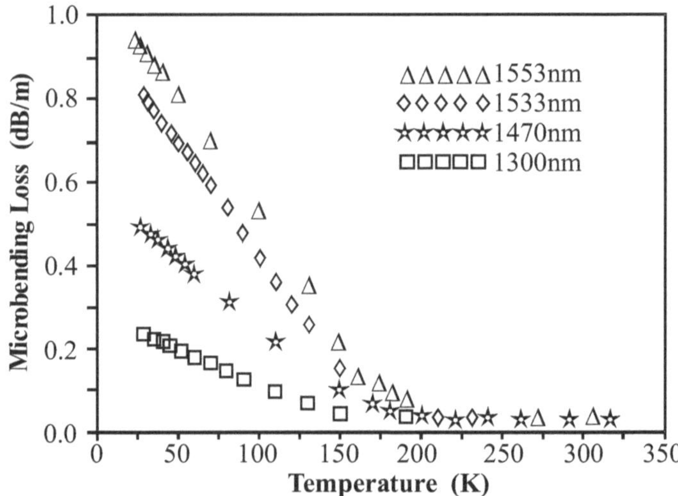

Figure 14.19: Microbending loss as a function of temperature for different wavelengths of the light source. Reprinted from ref. 4 with permission from OSA.

14.8.3 *Rare-earth doped optical fibers*

Rare-earth ions have many inner 4f-shell electron transitions that result in sharp absorption peaks in the visible and near infrared region of the spectrum. These rare earths can be easily incorporated in desired dopant concentrations into a variety of silica based glasses by MCVD processing. The absorption of light by these fibers depends on the temperature and hence if the light is launched into rare earth doped fiber the transmission will depend on the temperature of the fiber. The operation of a 2-5 cm long Nd^{3+} and Eu^{3+} rare earth doped fibers for temperature sensing was demonstrated a long time back[5]. Since then the temperature dependence of holomium[6], samarium[7] and praesodymium[8] have also been reported for sensor applications. Rare earth doped fibers can be used to sense temperatures in the range 0 to 100°C with 1°C resolution. Neodymium has been found to be most promising dopant for the temperature sensor.

14.8.4 *Liquid core fiber*

Liquid core optical fibers were investigated in 1972 to replace silica core fiber that showed high transmission losses[9,10]. As the transmission losses through pure silica fibers were brought down, research in liquid core fibers gone down. However, in 1983, hollow-core fibers filled with Kerr liquids were used for voltage measurements[11] followed it for a distributed temperature measurement[12]. Here we describe a temperature sensor that uses a hollow-core fiber filled with liquids with a well defined thermal response. It is based on the temperature dependent refractive index of the liquid and therefore similar to the sensor described earlier. The only difference is that there the liquid was used as a cladding and here it has been used as a core. In the previous sensor, increase in temperature makes the fiber highly guiding while in this case it makes the fiber weakly guiding. The liquid is chosen in such a way that it gives a significant drop in the intensity for the required temperature range. The liquid is filled in the hollow fiber using a vacuum process. The vacuum process can be accelerated by performing the filling of the fiber at elevated temperatures.

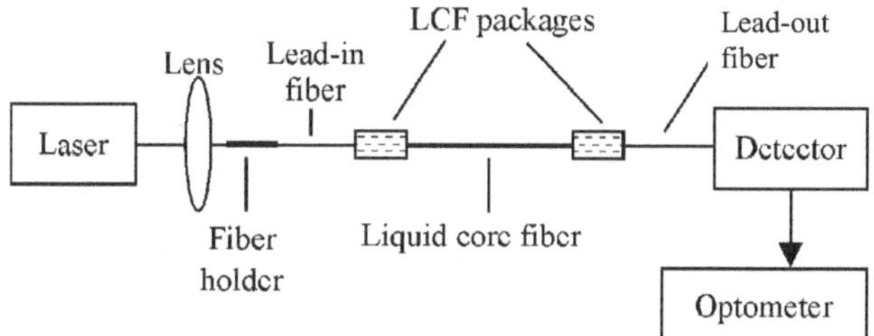

Figure 14.20: Liquid core fiber based temperature sensor.

The basic intensity based liquid core fiber set up is shown in Fig. 14.20. It consists of an optical source such as a He-Ne laser, collimated lens, liquid core fiber, a photodetector and lead fibers[13]. The LCF is surrounded by a silicone oil bath. The normalized transmitted intensity as a function of temperature of the silicone oil bath for a particular liquid filled is shown in Fig. 14.21. The intensity decreases sharply after it reaches 165°C. The range of the sensor is from 165°C to 185°C. By choosing liquids of different refractive indices different operating ranges can be obtained.

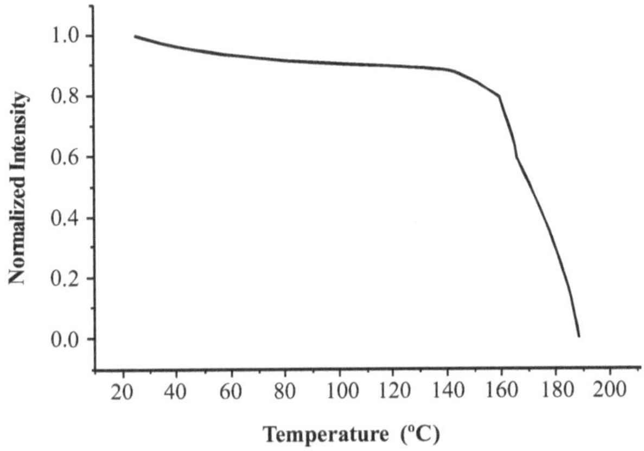

Figure 14.21: Typical variation of normalized transmitted intensity as a function of temperature.

All the temperature sensors described above are intrinsic type. Now we shall describe some extrinsic type temperature sensors. These are based on birefringence, thermochromic transducer, liquid crystal and thermal expansion properties of the metals.

14.8.5 Birefringent crystal

Birefringent crystals are optically transparent crystalline materials in which the indices of refraction are different for two perpendicularly polarized light waves. There are many birefringent crystals in which the index of refraction and hence the birefringence is a strong function of temperature. Figure 14.22 shows the sensor based on this property. It consists of two fibers, one to launch the light and the other to receive the light. The light is launched into the fiber and polarized before passing through the crystal. The light is then reflected from a mirror and passes back through the same crystal, polarizer and is coupled to the another fiber connected to a detector. As the temperature of the crystal changes the change in birefringence occurs. This changes the intensity of the light received which is proportional to the temperature. The sensor's range is limited to the biological process only.

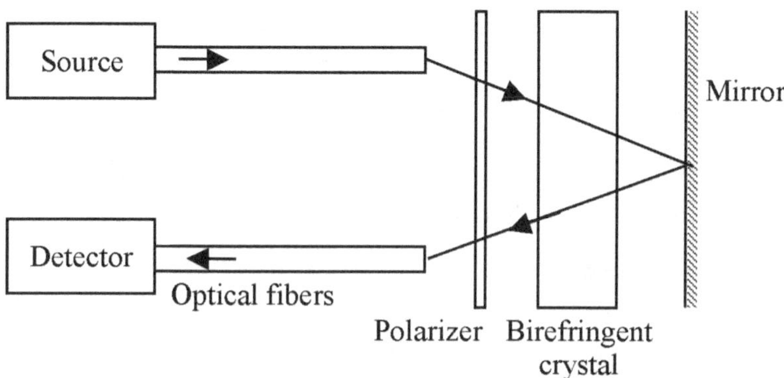

Figure 14.22: The temperature sensor based on the temperature dependent birefringence of a crystal.

14.8.6 Thermochromic transducer

There are a number of materials whose absorbance, at least at one wavelength, changes with temperature. These materials are called thermochromic. Various organic and inorganic materials possess this property. However in most of these materials, the change in absorbance with temperature is very small. Further, in some of the cases this characteristic is not reversible. Cobalt chloride is one of the thermochromic materials where the variation in absorbance with temperature is reversible. It is soluble in water, alcohols, acetone, ether etc. Its aqueous solution is pink to red, but turns blue when heated. Its colour is different in acetone or isopropyl alcohol. The solution of cobalt chloride in 15% water and 85% isopropyl alcohol absorbs light strongly at 660 nm while its absorbance at 840 nm is zero. Its absorbance at 660 nm is temperature dependent. As the temperature increases the absorbance increases.

A schematic diagram of the optoelectronic system constituting the sensor[14] is shown in Fig. 14.23. Light from a He-Ne laser is launched into two identical fibers of 200 μm core diameter. One fiber is used for the reference signal while the other for measuring signal. The measuring fiber enters the probe containing the thermochromic solution while the reference fiber is directly connected to the detector.

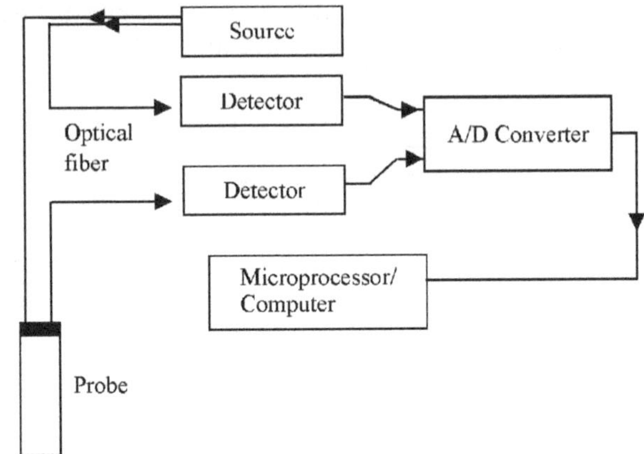

Figure 14.23: Schematic diagram of the optoelectronic system constituting thermochromic transducer optical fiber temperature sensor[14].

The light launched into the measuring fiber passes through the solution and is back reflected from the mirrored bottom of the probe. This light is then collected by the second fiber which is connected to another detector. The signals from the two detectors are fed into an A/D converter followed by a computer where the ratio of the two intensities is calculated. The ratio gives the temperature of the probe. In this design the reference fiber eliminates the source fluctuations. The sensor operates in the temperature range 25 to 60°C. The calibration curve of the sensor is shown in Fig. 14.24.

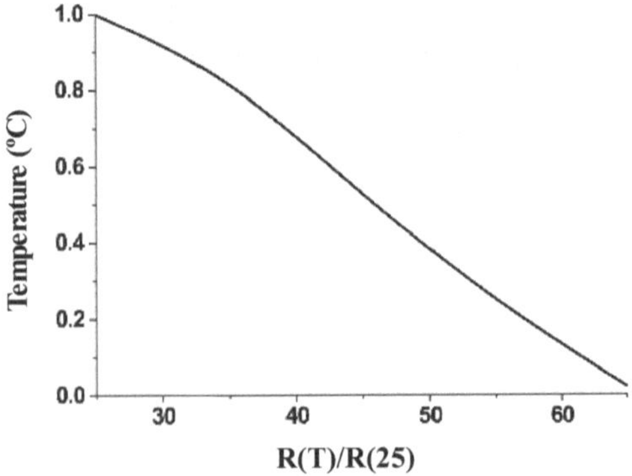

Figure 14.24: Typical calibration curve of the sensor[14].

14.8.7 Liquid crystals

Liquid crystal is a substance that share some of the properties of both liquid and crystal. Mechanically these substances resemble liquids in terms of flow, taking same shape as container and viscosity. They also possess the optical, electro-optical and magneto-optical properties of solid state crystals. A typical liquid crystal substance scatters light in symmetrical patterns and reflects different colours depending on the angle from which it is viewed. Depending on the arrangement of molecules in the crystal, liquid crystals are divided into three classes, smectic, nematic and cholesteric. The molecular structure of cholesteric liquid crystal is very delicately balanced and hence can be easily upset. Thus any small disturbance that interferes with the weak forces between the molecules can produce marked changes in optical properties such as reflection, transmission, birefringence, circular dichroism, optical activity and colour. The most striking optical transformation that occurs in a cholesteric liquid crystal in response to change in temperature is the change in colour. Although most cholesteric substances are colorless

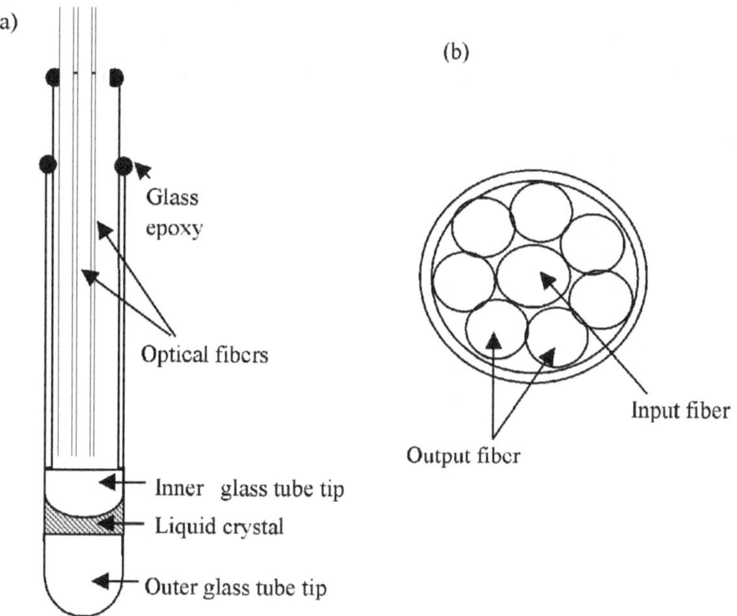

Figure 14.25: Liquid crystal optical fiber probe[15]. (a) Side view, (b) cross-sectional view.

liquids, they pass through a series of bright colours when they are cooled through their liquid-crystal phase. Most cholesteric liquid crystals pass through the colour phase within a temperature span of 2-3°C. However, the temperature span can be increased by mixing a number of liquid crystals in certain proportion. A fiber optic temperature sensor based on this property of the liquid crystals[15] is described here. The optoelectronic system of the sensor is the same as that shown in Fig. 14.23. The only difference is in the design of the probe. The probe consists of two concentric glass tubes with one end sealed and the other end opened Fig. 14.25 (a). The diameters of the two tubes are such that the inner diameter of the outer tube is equal to the outer diameter of the inner tube. A mixture of three cholesteric liquid crystals, namely, cholesteryl chloride (27.16%), cholesteryl nonanoate (43.98%) and cholesteryl carbonate (28.86%) is placed at the bottom between the two tubes. The inner tube contains seven multimode fibers as shown in the cross-section of the probe Fig. 14.25 (b). The light from the He-Ne laser is launched into the middle fiber. The reflected light from the liquid crystal is collected by the remaining six fibers which are connected to the detector. The rest of the procedure is the same as that in the case of the thermochromic transducer. The calibration curve of the sensor is shown in Fig. 14.26. The range of the sensor is from 34 to 42°C and hence can be used for biomedical applications.

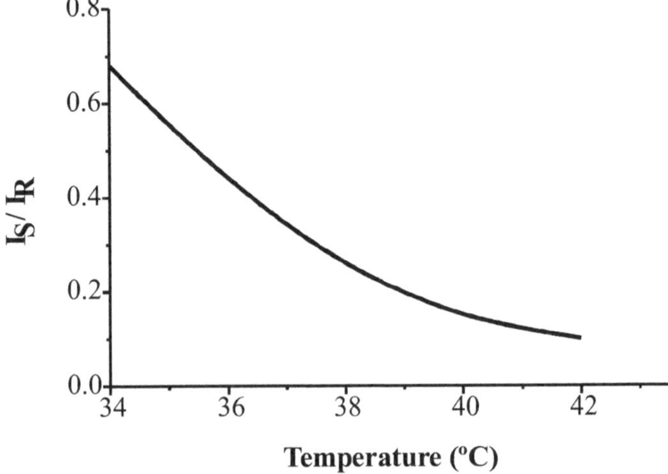

Figure 14.26: Calibration curve of the sensor based on liquid crystals[15].

14.8.8 Thermal expansion of metals

In this method the temperature dependent linear expansion of the metal is used. Different metals have different thermal coefficient of linear expansion. Figure 14.27 shows a temperature sensor that uses a bimetal strip formed by a copper and a steel strip of the same dimensions revetted together[16].

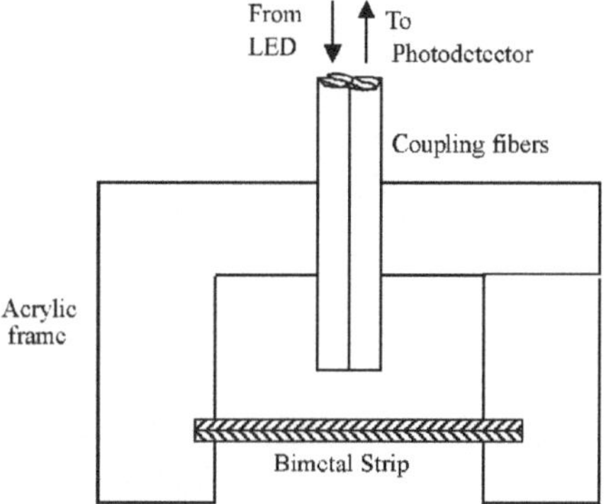

Figure 14.27: Schematic diagram of a fiber-optic temperature sensor using a bimetallic strip[16]

An acrylic frame holds the two ends of the bimetal strip and the two fibers. The fibers face the copper surface of the strip that also serves as a reflector. Light is launched into one fiber and is collected by the other fiber after reflection from the copper surface. The coefficients of thermal expansion of copper and steel are $18 \times 10^{-6}\ K^{-1}$ and $12 \times 10^{-6}\ K^{-1}$ respectively. If the temperature of the bimetal strip changes the strip bends and hence the light coupled to the receiving fiber alters. Thus by measuring the intensity detected one can find the temperature. The calibration curve of the sensor is shown in Fig. 14.28. Its dynamic range is from 35 to 135°C.

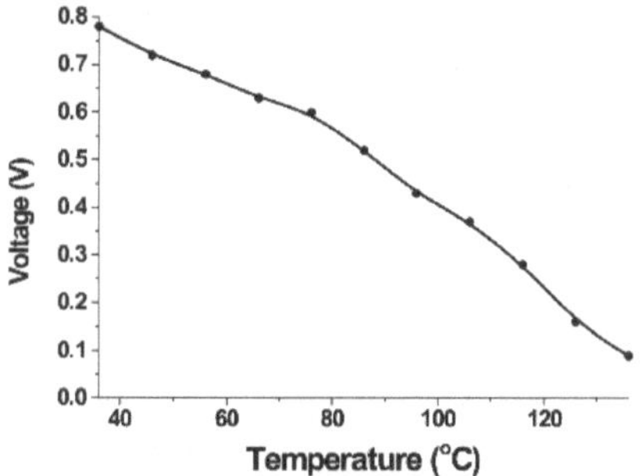

Figure 14.28: Calibration curve of the sensor[16].

14.9 Pressure sensor

Fiber optic sensors for pressure measurements have undergone extensive development. The industrial applications of these sensors include pressure monitoring in boilers, chemical process reactors and engines. In biomedical sciences, pressure measurements are widely performed in cardiology, urology and neurology. In fact, fiber optic sensors with microtip have been developed to measure pressures in arteries, bladder, urethra and rectum. Similar to temperature sensors, both intrinsic and extrinsic types pressure sensors have been developed. First we shall describe an intrinsic fiber optic pressure sensor based on microbending loss.

14.9.1 Microbending loss

Consider the propagation of rays in a step-index multimode optical fiber. Two rays with different angles of incidence are launched in the fiber as shown in Fig. 14.29. If the angles of these rays with the normal to the core cladding-interface inside the fiber are greater than or equal to the critical angle θ_c (= $\sin^{-1}(n_{cl}/n_1)$) of the fiber then these rays will confine in the core of the fiber. When the angle of any such ray becomes smaller than the critical angle, the ray exits the fiber from the cladding-air interface. Now let us bend the fiber. The ray which was guided in the straight region of the fiber may not remain guided in the bent region. This is because the angle of the ray with the normal to the core-cladding interface decreases in the bent region. If the new angle is smaller than the critical angle the ray will exit the fiber through the cladding. Figure 14.29 shows what happens to two rays launched in the fiber. When the ray with angle θ_2 is incident in the bent region its angle with the normal to the interface becomes θ'_2 which, of course, is smaller than θ_2. In the figure this angle has been assumed to be smaller than θ_c and therefore the ray exits the fiber. Another ray with angle θ_1 makes angle θ'_1 in the bent region which is smaller than θ_1 but greater than θ_c and hence it remains guided in the fiber. If the radius of curvature of the fiber is further decreased then the angle θ'_1 may become smaller than θ_c and, like θ_2 ray, this ray will also exit the fiber. Thus the power guided inside the fiber can be decreased by decreasing the bending radius of the fiber. In the pressure sensor, microbends are introduced in the fiber by using a deformer element. When the deformer element is displaced, for example, by applying an external pressure, mechanical deformation of the optical fiber perpendicular to its axis causes the higher order modes or the rays with angles close to θ_c to radiate out of the fiber core through the core-cladding interface. Thus the power of the light transmitted through the fiber will decrease as the pressure or the displacement is increased.

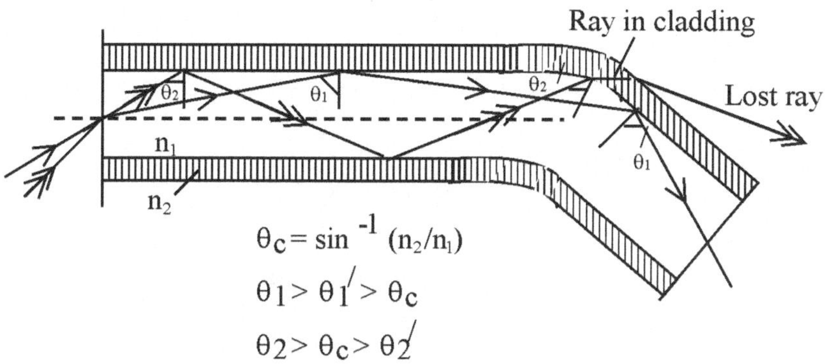

Figure 14.29: Propagation of rays in a bent fiber.

Figure 14.30 shows the basic configuration of the fiber optic microbend pressure sensor. Light is launched from a He-Ne laser using a microscope objective into a multimode optical fiber. A power meter is connected at the other end to measure the transmitted power.

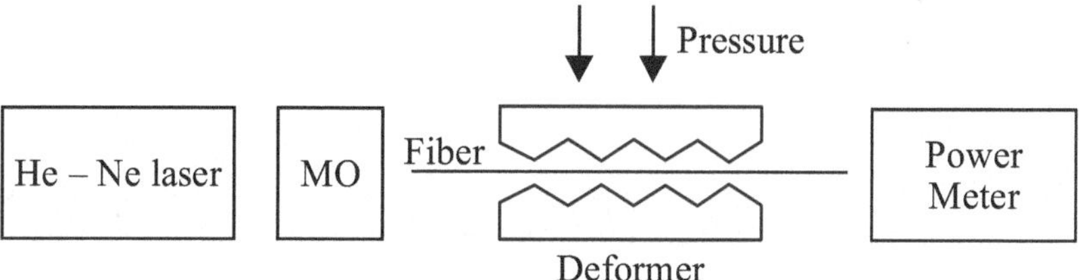

Figure 14.30: Schematic of a simple multimode optical fiber pressure sensor based on microbending.

The fiber is kept inside the deformer. As the pressure is applied (from the top) on the deformer, microbending is introduced along the length of the fiber, and this bending results in the loss of transmitted power by radiation at the bends. The loss in amplitude for a bent fiber is given by[17]

$$\text{loss} = C\left(\frac{a}{R}\right)^2 \qquad (14.6)$$

where R is the radius of curvature of the bend, a is the radius of the fiber core and C is a constant. Thus for a given fiber the pressure applied can be related to the bend radius; the bend radius is given by (see Fig. 14.31)

$$R = \frac{y^2 + D^2}{2y} \tag{14.7}$$

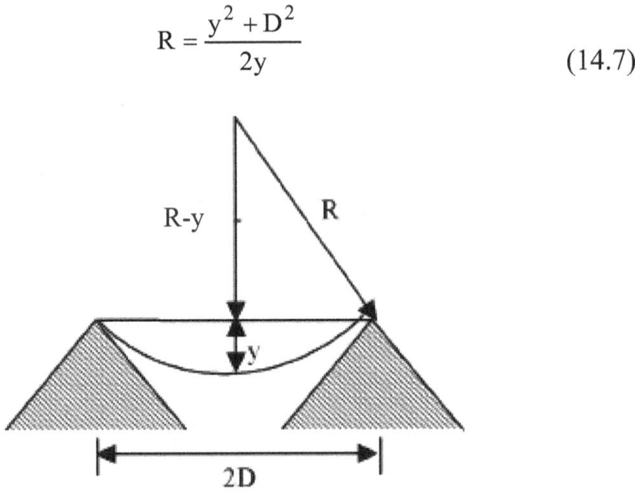

Figure 14.31: Geometry of the microbend.

where y is the displacement of the deformer element and 2D is the distance between the deformer element's contact points which is equal to the pitch of the deformer element. Thus the transmittance, T, through the fiber is

$$T = 1 - \text{loss} = 1 - \frac{Ca^2}{\left(\frac{y^2 + D^2}{2y}\right)^2} \tag{14.8}$$

$$T = 1 - C' \left(\frac{q}{1+q^2}\right)^2 \tag{14.9}$$

where

$$C' = \frac{4Ca^2}{D^2} \quad \text{and} \quad q = \frac{y}{D}$$

The applied force and hence the pressure is proportional to the displacement y. Therefore, in terms of pressure, we have

$$q = \frac{PA}{kD}$$

where P is the applied pressure, A is the surface area of the deformer and k is a constant. Figure 14.32 shows the variation of transmittance as a function of q for two different values of C'. The typical results obtained with graded index multimode fiber are shown in Fig. 14.33. It can be seen that the transmittance varies almost parabolically with the applied pressure as shown in Fig. 14.32.

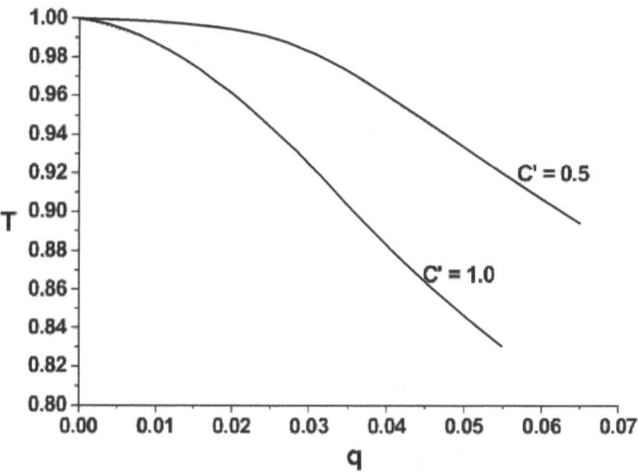

Figure 14.32: Theoretical variation of transmittance as a function of q.

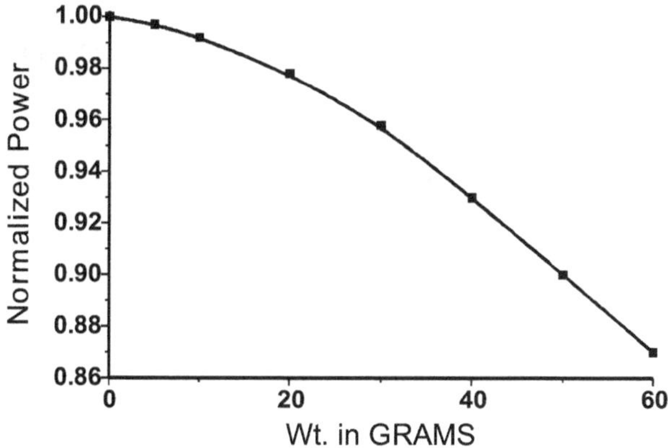

Figure 14.33: Typical plot of transmittance as a function of load on the deformer element for a multimode graded index fiber.

The pressure sensor described above is intrinsic type sensor. Now we describe extrinsic type pressure sensors.

14.9.2 Diaphragm curvature

In the sensor based on diaphragm curvature[18], light is brought to the diaphragm surface by a circle of fibers such that the light falls upon the diaphragm at a fixed radial distance from the centre of the diaphragm. The light reflected from the diaphragm is then distributed among collection fibers arranged concentrically with the illumination fibers both inside and outside see Fig. 14.34. The ratio of the intensities received by the outside fibers to the inside fibers is used to determine the pressure outside the diaphragm. The ratio method compensates for variation in the source intensity, losses in the input fibers and the variation in the reflectivity of the diaphragm surface. Figure 14.35 shows a side view of such a sensor. When the pressure outside the diaphragm is same as that inside, the diaphragm surface will be flat and an equal amount of light will be reflected to the inside and outside receiving fibers. If the pressure outside is more, the diaphragm will take the shape of a concave mirror and more light will be reflected to the outside collection fibers. Conversely, if the pressure outside is low then the surface will become convex and more light will be collected by inside fibers. Thus the pressure can be sensed.

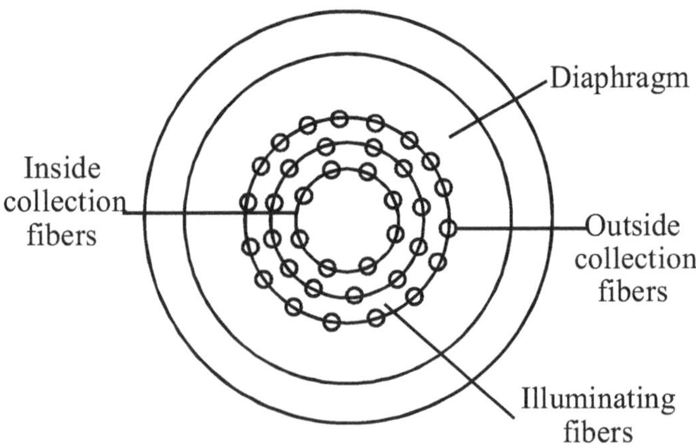

Figure 14.34: End view of a reflective diaphragm curvature probe[18].

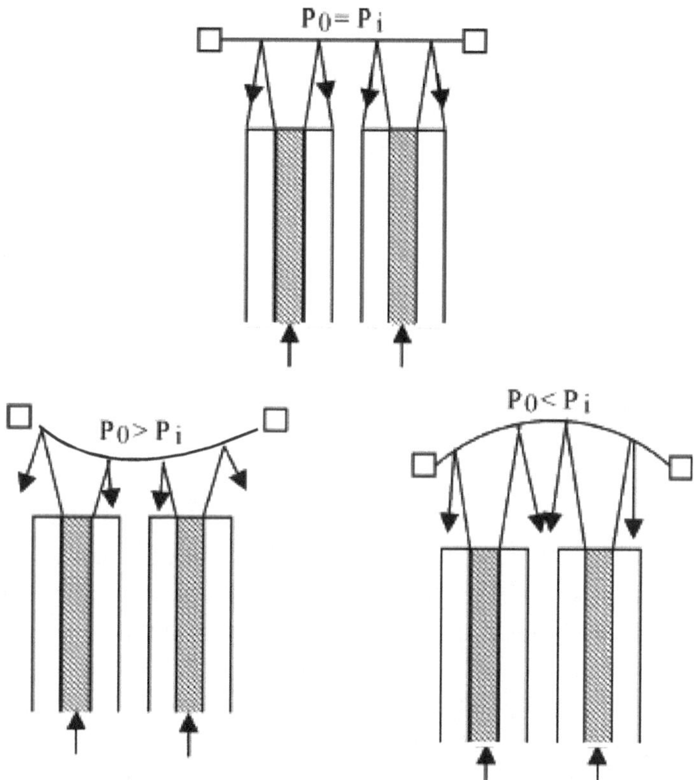

Figure 14.35: Side view of the pressure sensor[18].

14.9.3 Cantilever

In this sensor a cantilever balancing mirror suitably positioned at the end of a fiber bundle containing three fibers is used. The cantilevered mirror is attached to a membrane located on the side of the probe as shown in Fig. 14.36. If the outside pressure is different from the one that exists inside, the deflection of the membrane from the original position will occur. Thus the light exiting from the central fiber will reflect differentially toward either of the two light collecting fibers on each side of the central fiber. The light collected by these two pereferial fibers is sensed by an instrument which applies a feedback air pressure to the interior of the probe through a tube containing optical fibers to bring the membrane at its original position. The feedback pressure is thus a measure of the outside pressure.

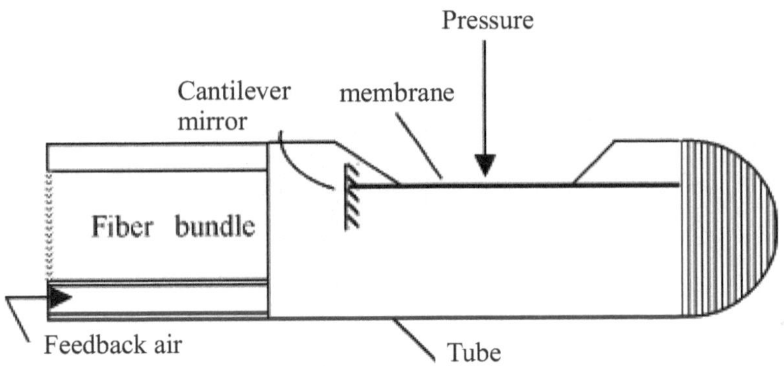

Figure 14.36: Fiber optic pressure sensor based on cantilever mirror.

14.10 Hydrocarbons detection in water

Chlorinated hydrocarbon solvents have been widely used in chemical and industrial processes, e.g. in the production of PVC, as solvents and as degreasing agents for metal surfaces. Large amounts of these substances have been spilled in the environment and are now contaminating soil, drainage and ground waters. Organochlorides, mainly trichloro- and tetra-chloroethylene, are major contaminants in ground, surface and drinking water and are mostly toxic or even carcinogenic. To assess the pollution of natural surroundings, the knowledge of the concentration of these pollutants is very important. There are several analytical techniques (chromatographic methods) to quantify these contaminants.

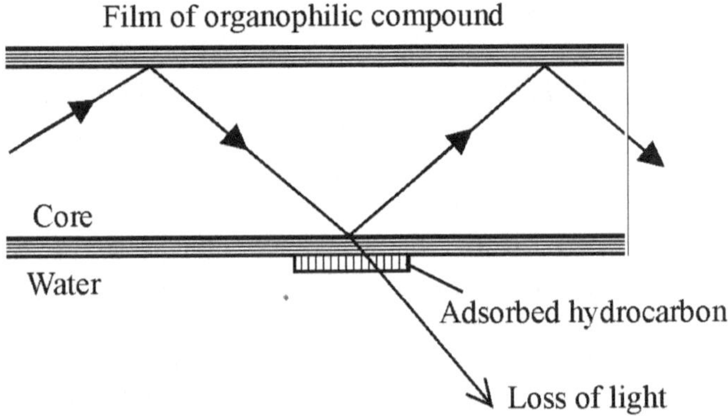

Figure 14.37: Detection of hydrocarbons in water.

However these techniques are not very sensitive. Further these are not suitable for in situ or on site measurements. Optical fiber sensors have been developed to remove these disadvantages. The first fiber-optic hydrocarbon sensor for fresh and salt water was developed based on the change in refractive index of the fiber cladding[19]. In the sensor, unclad fiber is coated with an organophilic compound that adsorbs hydrocarbons. Because the index of refraction of water is lower than the fiber core, total internal reflection occurs with no light loss. When a small quantity of hydrocarbon comes in contact with the coated unclad fiber surface the resultant higher refractive index causes the light to refract out of the core Fig. 14.37. This reduces the light intensity at the detector. This light loss is then related to the quantity of the adsorbed contaminant.

14.11 Oxyhemoglobin concentration measurements

Optical fibers can be used to determine directly the oxygen content in the blood. In blood, there are two kinds of hemoglobin molecules, one carrying oxygen and the other not carrying it. The oxyhemoglobin is important for our life because the oxygen carrying by it is used for the functions of various kinds of cells in the body. The amount of oxygen carried by the hemoglobin in erythrocytes relative to the maximum carrying capacity is called oxygen saturation. The arterial blood is about 95% saturated while the venous blood may be 75% saturated. The increase of PO_2 (partial pressure of oxygen) or pH of the blood generally causes an increase of oxyhemoglobin saturation while increase of PCO_2 or temperature reduces oxyhemoglobin saturation. Optical fibers can be used to monitor the whole oxygen saturation. The absorption spectra of hemoglobin and oxyhemoglobin are shown in Fig. 14.38. It can be seen that oxyhemoglobin reflects red light more than hemoglobin. In place of red light if infrared light is used then all the hemoglobin whether carrying oxygen or not reflect light equally. Thus dual wavelength spectrometry can be performed on the blood to determine the concentration of oxyhemoglobin Fig. 14.39. Two sources of light, one emitting red (660 nm) and the other emitting infrared (805 nm) are used. The device used was a 1x3 fiber coupler fitted into a catheter in which the 660 nm wavelength light is coupled to one fiber while 805 nm wavelength light is sent through the other fiber. Lights at these two wavelengths exiting from the tip of the coupler are reflected by the hemoglobins present in the blood. The reflected light is collected by the third fiber whose end is connected to the detector. The intensity of the reflected red light gives the amount of oxyhemoglobin whereas the intensity of the reflected infrared light determines the total amount of hemoglobin in the blood. The ratio of two intensities I(660)/I(805) determines the oxyhemoglobin concentration. From this device it is possible to determine oxyhemoglobin concentration in the range 60 to 90%.

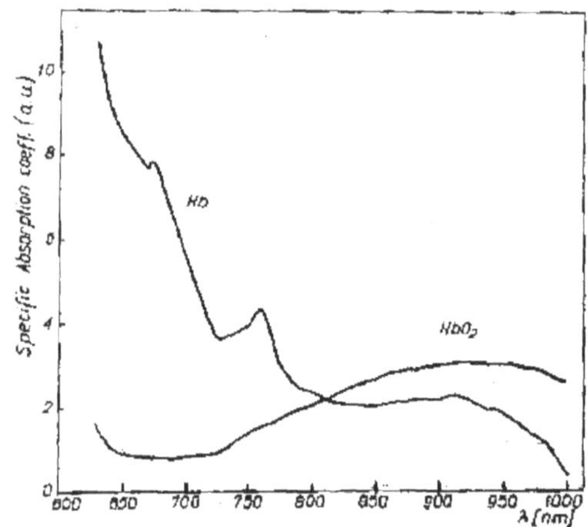

Figure 14.38: Absorption spectra of hemoglobin (Hb) and oxyhemoglobin (HbO_2).

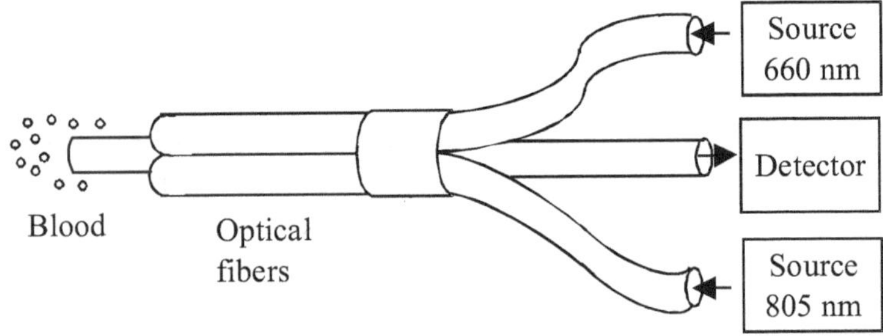

Figure 14.39: Blood oximetry by reflectance analysis.

The above mentioned technique is very important for long term monitoring of blood oxygenation while treating cardiac surgery patients. The advantage is that there is no need of repeated sample taking to determine the oxygen content. From this technique one can also get information whether the heart and lungs are properly functioning or not in terms of oxygen supply.

14.12 Cardiac output measurement

The above technique can also be used to calculate the cardiac output. For this a dye, called indocyanine green, is injected in the blood. The dye has an absorption peak at 805 nm while its absorption at wavelengths greater than 900 nm is zero. These two wavelengths are used in 1x3 coupler (Fig. 14.39). The light of wavelength 805 nm is absorbed by the dye in proportion to its concentration in the blood while the light of wavelength 900 nm is not absorbed by the dye. The ratio of two light intensities collected by the third fiber gives the concentration of the dye. The time variation of the dye concentration is used to calculate the cardiac output.

In this chapter we have described a selected number of intensity modulated fiber optic sensors for the measurement of physical, chemical and biological parameters. In some of the intensity modulated sensors, loss in intensity occurs due to connections and splices, microbending, mechanical creep and misalignment of light sources and detectors due to environmental effect. Further, the fluctuations in light source can affect performance of the sensor. To overcome these, one needs referencing to cancel out these losses.

References

1. D.C. O'Shea (1985) Elements of Modern Optical Designs. Wiley, New York, p.2 59.
2. A. Kumar, T.V.B. Subrahmanyam, A.D. Sharma, K. Thyagarajan, B.P. Pal and I.C. Goyal (1984) Novel refractometer using a tapered optical fibre. *Electron. Lett.* **20**, 534-535.
3. A.M. Scheggi, M. Brenci, G. Conforti, R. Falciai and G.P. Preti (1984) Optical fiber thermometer for medical use. *Proc. SPIE* **494**, 13-17.
4. O. Lumholt, A. Bjarklev, S.D. Petersen, C.C. Larsen, J.H. Povlsen, T. Rasmussen and K. Rottwitt (1991) Simple fiber-optic low temperature sensor that uses microbending loss. *Opt. Lett.* **16**, 1355-1357.
5. E. Snitzer, W.W. Morey and W.H. Glenn (1983) Fiber optic rare-earth temperature sensors. *Proc. Ist Int. Conf. on Optical Fiber Sensors*, pp. 79-82.

6. M.C. Farries (1987) Distributed temperature sensor using holmium doped fiber, *Proc. 6th Int. Conf. on Integrat. Opt. and Opt. Fiber Commun.* (Reno, NV).
7. M.C. Farries and M.E.Fermann (1987) Temperature sensing by thermally induced absorption in neodymium doped optical fiber. *Proc. SPIE* **798**, 115-120.
8. K.W. Quoi, R.A. Lieberman, L.G. Cohen, D.S. Shenk and J.R. Simpson (1992) Rare-earth doped optical fibers for temperature sensing. *J. Lightwave Technol.* **10**, 847-852.
9. J. Stone (1972) Optical transmission in liquid-core quartz fibers. *Appl. Phys. Lett.* **20**, 239-240.
10. D.N. Payne and W.A. Gambling (1972) New low-loss liquid core fiber waveguide. *Electron. Lett.* **8**, 374-376.
11. M. Kuribara and Y. Takeda (1983) Liquid core optical fiber for voltage measurement using Kerr effect. *Electron. Lett.* **19**, 133-135.
12. A.H. Hartog (1983) A distributed temperature sensor based on liquid core optical fibers. *J. Lightwave Technol.* **1**, 498-509.
13. M. deVries, B.D. Zimmermann, A.M. Vengsarkar and R.O. Claus (1991) Liquid core optical fiber temperature sensor. *IEEE Region 3 Technical Conference*, Huntsville (USA).
14. B.D. Gupta, A. Sharma and S.K.S. Nair (1992) Fiber-optic temperature sensor for biomedical applications. *J.I.E.T.E.* **38**, 368-370.
15. B.D. Gupta, A. Sharma and R. Goyal (1993) Optical fiber temperature sensor based on liquid crystals. *Int. J. Optoelectron.* **8**, 13-19.
16. K.S. Lau, W.H. Wong and S.K. Yeung (1992) Fiber optic sensors for laboratory measurements. *Eur. J. Phys.* **13**, 227-235.
17. C.K. Kao (1982) Optical fiber systems: technology, design and applications. McGraw-Hill, New York, p.41.
18. C.M. Lawson and V.J. Tekippe (1983) Fiber optic diaphragm-curvature pressure transducer. *Opt. Lett.* **8**, 286-288.
19. F.K. Kawahara, R.A. Fiutem, K.S. Silvus, F.M. Newman and J.K.Frazar (1983) Development of novel method for monitoring oils in water. *Anal. Chim. Acta* **151**, 315-327.
20. J.M. Schmitt, F.G. Mihm and J.D. Meindl (1986) New methods for whole blood oximetry. *Annals Biomed. Engn.* **14**, 35-52.

Index

T

U

V

W

Zeitfracht Medien GmbH
Ferdinand-Jühlke-Straße 7
99095 Erfurt, Deutschland
produktsicherheit@kolibri360.de